Fuzzy Logic Dynamics and Machine Prediction for Failure Analysis

Tawanda Mushiri
University of Johannesburg, South Africa

Charles Mbowhwa
University of Johannesburg, South Africa

A volume in the Advances
in Computer and Electrical
Engineering (ACEE) Book Series

Published in the United States of America by
    IGI Global
    Engineering Science Reference (an imprint of IGI Global)
    701 E. Chocolate Avenue
    Hershey PA, USA 17033
    Tel: 717-533-8845
    Fax: 717-533-8661
    E-mail: cust@igi-global.com
    Web site: http://www.igi-global.com

Library of Congress Cataloging-in-Publication Data

Names: Mushiri, Tawanda, author. | Mbohwa, Charles, author.
Title: Fuzzy logic dynamics and machine prediction for failure analysis / by
  Tawanda Mushiri and Charles Mbohwa.
Description: Hershey PA : Engineering Science Reference, [2018] | Includes
  bibliographical references.
Identifiers: LCCN 2017015199| ISBN 9781522532446 (hardcover) | ISBN
  9781522532453 (ebook)
Subjects: LCSH: Machinery--Monitoring. | Fuzzy logic. | Failure analysis
  (Engineering)
Classification: LCC TJ153 .M798 2018 | DDC 621.8028/7--dc23 LC record available at https://
lccn.loc.gov/2017015199

This book is published in the IGI Global book series Advances in Computer and Electrical Engineering (ACEE) (ISSN: 2327-039X; eISSN: 2327-0403)

British Cataloguing in Publication Data
A Cataloguing in Publication record for this book is available from the British Library.

All work contributed to this book is new, previously-unpublished material.
The views expressed in this book are those of the authors, but not necessarily of the publisher.

For electronic access to this publication, please contact: eresources@igi-global.com.

Advances in Computer and Electrical Engineering (ACEE) Book Series

ISSN:2327-039X
EISSN:2327-0403

Editor-in-Chief: Srikanta Patnaik, SOA University, India

MISSION

The fields of computer engineering and electrical engineering encompass a broad range of interdisciplinary topics allowing for expansive research developments across multiple fields. Research in these areas continues to develop and become increasingly important as computer and electrical systems have become an integral part of everyday life.

The **Advances in Computer and Electrical Engineering (ACEE) Book Series** aims to publish research on diverse topics pertaining to computer engineering and electrical engineering. **ACEE** encourages scholarly discourse on the latest applications, tools, and methodologies being implemented in the field for the design and development of computer and electrical systems.

COVERAGE

- VLSI Design
- Sensor Technologies
- Programming
- Computer science
- Circuit Analysis
- Microprocessor Design
- Optical Electronics
- Power Electronics
- Electrical Power Conversion
- Applied Electromagnetics

IGI Global is currently accepting manuscripts for publication within this series. To submit a proposal for a volume in this series, please contact our Acquisition Editors at Acquisitions@igi-global.com or visit: http://www.igi-global.com/publish/.

The Advances in Computer and Electrical Engineering (ACEE) Book Series (ISSN 2327-039X) is published by IGI Global, 701 E. Chocolate Avenue, Hershey, PA 17033-1240, USA, www.igi-global.com. This series is composed of titles available for purchase individually; each title is edited to be contextually exclusive from any other title within the series. For pricing and ordering information please visit http://www.igi-global.com/book-series/advances-computer-electrical-engineering/73675. Postmaster: Send all address changes to above address. ©© 2018 IGI Global. All rights, including translation in other languages reserved by the publisher. No part of this series may be reproduced or used in any form or by any means – graphics, electronic, or mechanical, including photocopying, recording, taping, or information and retrieval systems – without written permission from the publisher, except for non commercial, educational use, including classroom teaching purposes. The views expressed in this series are those of the authors, but not necessarily of IGI Global.

Titles in this Series

For a list of additional titles in this series, please visit:
https://www.igi-global.com/book-series/advances-computer-electrical-engineering/73675

Free and Open Source Software in Modern Data Science and Business Intelligence...
K.G. Srinivasa (CBP Government Engineering College, India) Ganesh Chandra Deka (M. S. Ramaiah Institute of Technology, India) and Krishnaraj P.M. (M. S. Ramaiah Institute of Technoloy, India)
Engineering Science Reference • ©2018 • 189pp • H/C (ISBN: 9781522537076) • US $190.00

Design Parameters of Electrical Network Grounding Systems
Osama El-Sayed Gouda (Cairo University, Egypt)
Engineering Science Reference • ©2018 • 316pp • H/C (ISBN: 9781522538530) • US $235.00

Design and Use of Virtualization Technology in Cloud Computing
Prashanta Kumar Das (Government Industrial Training Institute Dhansiri, India) and Ganesh Chandra Deka (Government of India, India)
Engineering Science Reference • ©2018 • 315pp • H/C (ISBN: 9781522527855) • US $235.00

Smart Grid Test Bed Using OPNET and Power Line Communication
Jun-Ho Huh (Catholic University of Pusan, South Korea)
Engineering Science Reference • ©2018 • 425pp • H/C (ISBN: 9781522527763) • US $225.00

Transport of Information-Carriers in Semiconductors and Nanodevices
Muhammad El-Saba (Ain-Shams University, Egypt)
Engineering Science Reference • ©2017 • 677pp • H/C (ISBN: 9781522523123) • US $225.00

Accelerating the Discovery of New Dielectric Properties in Polymer Insulation
Boxue Du (Tianjin University, China)
Engineering Science Reference • ©2017 • 388pp • H/C (ISBN: 9781522523093) • US $210.00

Handbook of Research on Nanoelectronic Sensor Modeling and Applications
Mohammad Taghi Ahmadi (Urmia University, Iran) Razali Ismail (Universiti Teknologi Malaysia, Malaysia) and Sohail Anwar (Penn State University, USA)
Engineering Science Reference • ©2017 • 579pp • H/C (ISBN: 9781522507369) • US $245.00

For an entire list of titles in this series, please visit:
https://www.igi-global.com/book-series/advances-computer-electrical-engineering/73675

701 East Chocolate Avenue, Hershey, PA 17033, USA
Tel: 717-533-8845 x100 • Fax: 717-533-8661
E-Mail: cust@igi-global.com • www.igi-global.com

Table of Contents

Preface

INTRODUCTION

This book analyses machinery failures using fuzzy logic to monitor different parameters. This research monograph presents research results and studies of six companies in a developing country based on fuzzy logic approaches and models to machinery failure analysis. The major objective of the study was simulating and modelling machinery and then coming up with a suitable intelligent Condition Based Maintenance system; to do a root cause analysis on the bottle washer and apply fuzzy logic with Model Reference Adaptive Control to prevent failure to the Beverages Production Company bottle washer; introduce an intelligent CBM program; find ways to avoid clinker formation and identify causes of clinker formation on platen super heater elements for the thermal power generation company; to determine the causes of failure of the cone crusher, estimate time between major crusher failures and provide suitable action plan; optimization of the crusher circuit for the Platinum Mining Company; coming up with a control system that monitors water in the dam as well as protecting the equipment from failure and do stability control of machinery; to develop a Computerized Maintenance Management System prototype using MS Access; to develop a real-time artificial intelligent plant machinery online monitoring system for the Fruits Processor Company; apply the concept of Maintenance Free Operating Period (MFOP) reliability metric to determine maintenance intervals as opposed to Mean Time Between Failure (MTBF) to the Hydro-Power Generation Company and also do a Simulink control on the governor. The results of the simulation of the control strategy of Fuzzy Logic Proportional Integral Differential have much preferred performance as compared to the general Proportional Integral Differential Controller. This was done in the Simulink/Matlab simulation environment. The work addresses problems resulting from the failure to monitor nonlinear systems that naturally exist in such environments. Examples of specific work and contributions of this research work are discussed.

Challenges

The main problem was found to be resulting from failure to monitor nonlinear systems that naturally exist in such environments. When the complexity of the controller plant is a non-linear system, the production process cannot be easily maintained optimally and it is not easy to manipulate the parameters on Proportional Integral Differential Controller.

Solutions

Generalized integrated fuzzy condition-based maintenance systems and procedures were developed incorporating programmable logic control, supervisory control and data acquisition for monitoring plant machinery. This is a unique methodology that has provided new approaches and insights in this domain. A major contribution of this work is the demonstration that model reference adaptive fuzzy controller has better performance than model reference adaptive controller that is coupled with a proportional integral differentiator controller when applied to condition-based maintenance. This book has developed systems and recommendations that can be used by researchers and academics in the application of fuzzy logic and systems to condition-based maintenance to real world systems. Contributions from research work done at a Platinum Mining Company was the application of root cause analysis technique to determine the major causes of failure of the pebble crusher, within the context of a company operation in a developing country. The analysis of stresses was done using solid works software and condition monitoring was done using Matlab 2015 software to note the development of the cracks in a shaft. At the Beverages Manufacturing Company, failure analysis was done and significant loss of production in the plant was reduced. A model reference adaptive fuzzy controller was designed for the pneumatic valve. The error resulting from the difference between the actual system output and that of the reference model was analysed using a fuzzy logic controller. At Fruits Processor Company, a beverage industry company; a new and better maintenance strategy was proposed based on our research results. A fuzzy logic controller was designed for the critical machinery using the fuzzy logic toolbox in Matlab. Following the failure analysis, a Computerized Maintenance Management Systems program was designed using Ms Access and Ms Excel tools to improve overall plant performance. A Programmable Logic Controller Ladder program was also designed for this maintenance program. At a Hydro Power Generation Company the administering system of a 150

MW Francis Turbine was developed using fuzzy logic. The representative framework parameters were mirrored with the real information accessible in the power plant. At Thermal Power Distribution Company, the main control parameters in the boiler were modelled. These included coal flows, temperatures in the combustion zone, air to fuel ratio, ash content that is the percentage of ash in raw coal, and mineral content. These parameters were optimised in order to avoid clinker formation on the platen super heater tubes in the boiler. A Condition-Based Maintenance strategy and fuzzy logic were used to monitor the boiler parameters. At a Water Distribution Company the problem of corroding gates and failure was solved using solid works, Matlab and Programmable Logic Controller. Simulink was used to monitor online to avoid further failure. The results of these studies can be used to inform research work and decisions on maintenance strategy by companies in different fields in both developing ad industrialized economies.

ORGANIZATION OF THE BOOK

The book is organized into 10 chapters. A brief description of each of the chapters follows:

Chapter 1

Chapter 1 introduces the whole book with fuzzy logic usage in machinery combined with condition based maintenance. One of the most important tools in minimizing downtime, whether or not a conventional preventive maintenance program is possible, is called "preventive engineering." Although this would appear to be the application of common sense to equipment design maintenance engineering, it is a field which is often neglected. Too often maintenance engineers are so busy handling emergency repairs or in other day-to-day activities that they find no opportunity to analyze the causes for breakdowns which keep them so fully occupied. While most engineers keep their eyes open to details such as better packings, longer-wearing bearings, and improved lubrication systems, true preventive engineering goes further than this and consists of actually setting aside a specific amount of technical manpower to analyze incidents of breakdown and determine where the real effort is needed; then through redesign, substitution, changes, and specifications, or other similar means, reducing the frequency of failure and the cost of repair (Linderly, Higgins, & Mobley, 2002).

Chapter 2

Chapter 2 summarises the need for condition based maintenance. It was noted that maintenance is centered on world war two; before, during and after the war. This accelerated many developments in machinery monitoring

Chapter 3

Chapter 3 explains where fuzzy logic is currently being used. Fuzzy has gained a lot of ground as it is used in fault diagnosis (Mustapha, Sapun, Ismail, & Mokhtar, 2004), structural reliability (Savoia, 2002; Wang, Pan, & Chen, 2006), human reliability (Konstandinidou, Nivolianitou, Kiranoudis, & Markatos, 2006), safety and risk engineering (Mostafa, Payam, Hatami, Alireza, & Allah, 2013; Guimaraes & Lapa, 2007), and quality control (Khan & Hafiz, 1999; Sharma, Kumar, & Kumar, 2007). This is what made the author to do the research using fuzzy logic. Kumar, Sharma, and Kumar (2007) developed a system that may help maintenance engineers to analyze and predict the system behavior in robots. An attempt has also been made to deal with imprecise, uncertain dependent information related to system performance. Various reliability parameters (such as failure rate, repair time, mean time between failures, availability, reliability and expected number of failures) were computed to predict the system behavior in objective terms and it is concluded that in order to improve the availability and reliability aspects, it is necessary to enhance the maintainability requirement of the system.

Chapter 4

Chapter 4 is based on fuzzy logic approaches to machinery failure analysis. Six companies and organisations were used as model validation case studies. At a platinum mining company, the research was based on the root cause analysis technique. The objective was to determine the major causes of failure of the pebble crusher, to estimate between the major crusher failures and provide suitable solutions that included the optimization of the crushing circuit. The second application was done on a beverages manufacturing companies focusing on the bottle washing process. The third company used a reactive and firefighting maintenance strategy which resulted in frequent catastrophic breakdowns, ever increasing maintenance costs and long unplanned plant shutdowns. The fourth was done for a hydropower generation company. The dynamic characteristics of these systems are nonlinear and difficult to predict. This fifth case study application focused on the water gate control and avoidance of its failure. The sixth and final case study application was at a thermal power generation company.

Chapter 5

Chapter 5 was on was done on a beverages manufacturing companies focusing on the bottle washing process. The main problem is that the pneumatic valve of the bottle washer which controls the discharge of clean bottles sometimes sticks or fails which results in significant loss of production since this is a bottleneck operation. The main causes of failure were found to be the temperature and pressure, which often fell outside the required ranges; and minor contributions to failure due to moisture and abrasive particles. In order to solve this problem a model reference adaptive fuzzy controller was designed for the pneumatic valve using the MATLAB software. The model reference adaptive control (MRAC) system consists of the reference model which has the desired output of the system. The error resulting from the difference between the actual system output and that of the reference model is executed by the fuzzy logic controller (FLC). The simulation of the behaviour of the valve in response to the reference model was done using Simulink.

Chapter 6

Chapter 6 presents a platinum mining company, the research was based on the root cause analysis technique. The objective was to determine the major causes of failure of the pebble crusher, to estimate between the major crusher failures and provide suitable solutions that included the optimization of the crushing circuit. Major failures were investigated including the breaking of the main shaft, bearing failure and also entry of tramp iron in the crushing chamber. In solving these problems analysis of stresses was done using solid works 2015, and condition monitoring techniques were applied using MATLAB 2015 to investigate the development of the crack in the shaft. The results showed that EN 19 has better physical properties than EN 9 and EN 26. EN 19 was recommended for the construction of the main shaft. Crack detection prediction by using MATLAB can be complemented and validated by the use of non-destructive testing.

Chapter 7

Chapter 7 was on Hydropower Generation Company; the focus was on making use of the Reliability Centred Maintenance (RCM) principles in relation to Expert Systems in order to optimize maintenance of power generation assets at HPGC station. The researchers realised that in order for CBM to come out clearly, it is critical to do the RCM first. Naturally, the ageing equipment demands a paradigm shift in maintenance strategies in order to guarantee continuity of supply and meet the ever-growing stakeholder requirements. Increase in customer demand has worsened the situation. There is need to improve the Overall Equipment Effectiveness (OEE) from the current 60% to the world class 85%, using the same old equipment in order to retain customer satisfaction.

Chapter 8

In Chapter 8, the research on gate control work presents the concept of regulating the flow of water by controlling the dam gates or shutters using servo-motors in order to manage the dam water level. Water level control and insufficient water supply are the major challenges at the hydropower company HDPC. The system at this dam site consists of the outlet works tower with four outlet penstocks that collects water to the main discharge outlet. The automation implements the electronic control system that uses the programmable logic controller (PLC) and the optical level sensors to detect water level. For the new design, the power screw hoist mechanism is used to open and close the gate of weight 600N with the efficiency of 35.4% of the square thread screw. The gate takes 160 seconds to travel a vertical distance of 600mm in its guide, to fully open and close the 400mm penstock diameter. Mat-lab Simulink was used to control the electronic system for stability to avoid vibrations and fluctuation of speed.

Chapter 9

Chapter 9 is the final case study application, at a thermal power generation company. In large scale industrial applications, the controlling and optimization of the parameters must be done efficiently and effectively so as to attain smooth operation of the plant. In this research the main control parameters in the boiler were identified as coal flows, temperatures in the combustion zone, air to fuel ratio and ash content, that is the percentage of ash in raw coal, and mineral content. These parameters are monitored to avoid clinker formation in the super heater tubes of the boiler. Condition-Based Maintenance (CBM) approaches were used to monitor the boiler parameters. Fuzzy logic was applied in monitoring of these parameters.

Chapter 10

In Chapter 10, the authors highly appreciated the use of fuzzy logic in machinery. Model Reference Adaptive Control and Model Reference Adaptive Fuzzy Control need to be worked on in future in mining machinery beverages, electricity generation since they are more accurate than the general PLCs used in industries world-wide. Further research should be carried out using other techniques of AI to control valve operation and other components of the machine which contribute to the machine breakdown. These include the Genetic Algorithms (GA), Neural Networks (NN), Expert Systems (ES) and, Neuro Fuzzy (NF) systems. These techniques should each be tested on how they will respond to the bottle washer system changes and how they will be able to control the system processes. The possibility of changing the material of the valve to suit the conditions in the bottle washer should also be considered. Due to several limitations which include time, the subject under discussion was not exhausted and there is room for future studies and improvements on the same topic. Reliability Centred Maintenance Expert systems can be integrated with other modern intelligent systems like the neural networks and genetic algorithms to help improve their efficiency as well as design towards autonomous maintenance systems which would carry out certain maintenance tasks.

REFERENCES

Guimaraes, A. C. F., & Lapa, C. M. F. (2007). Fuzzy inference to risk assessment on nuclear engineering systems. *Applied Soft Computing*, *7*(1), 17–28. doi:10.1016/j.asoc.2005.06.002

Khan, M. K., & Hafiz, N. (1999). Development of an expert system for implementation of ISO quality systems. *Total Quality Management*, *10*(1), 47–59. doi:10.1080/0954412998054

Konstandinidou, M., Nivolianitou, Z., Kiranoudis, C., & Markatos, N. (2006). A Fuzzy modeling application of CREAM methodology for human reliability analysis. *Reliability Engineering & System Safety*, *91*(6), 706–716. doi:10.1016/j.ress.2005.06.002

Kumar, A., Sharma, S. P., & Kumar, D. (2007). *Robot reliability using petri nets and fuzzy lambda-tau methodology*. Reliability and Safety Engineering.

Linderly, R., Higgins, R., & Mobley, K. (2002). *Maintenance Engineering Handbook* (6th ed.). McGraw Hill.

Mostafa, M., Payam, M., Hatami, A. M., Alireza, G., & Allah, K. H. (2013). A Study of Barriers and Success Keys to The Implementation of Computerized Maintenance Management System in an Organization: Case Study in Fan Avaran Petrochemical Company. *Life Science Journal*, 20–34.

Mustapha, F., Sapun, S. M., Ismail, N., & Mokhtar, A. (2004). A computer based intelligent system for fault diagnosis of an aircraft engine. *Engineering Computation*, *21*(1), 78–90. doi:10.1108/02644400410511855

Savoia. (2002). Structural reliability analysis through fuzzy number approach, with application to stability. *Computers & Structures,* 1087–1102.

Sharma, R. K., Kumar, D., & Kumar, P. (2007). Quality costing in process industries through QCAS: A practical case. *International Journal of Production Research*, *45*(15), 3381–3403. doi:10.1080/00207540600774067

Wang, W. L., Pan, D., & Chen, H. M. (2006). Architecture-based software reliability modeling. *Journal of Systems and Software*, *79*(1), 132–146. doi:10.1016/j.jss.2005.09.004

Chapter 1
A New Technology of Fuzzy Logic in Machinery Monitoring

ABSTRACT

This chapter is based on fuzzy logic approaches to machinery failure analysis. Six companies and organisations were used as models for validation case studies. An introduction to fuzzy logic is done in this chapter of the book. As part of this book, it explains the marriage of condition-based maintenance (CBM) in artificial intelligence (AI) specifically using fuzzy logic as a subset of machine intelligence. The idea is to make the machine think intelligently like a human being. The International Standardization Organisation (ISO) has now set up a committee of condition monitoring and diagnosis of machines. Many manufacturing companies are pushing their production equipment for every ounce of capacity while, at the same time, trying to cut their overhead costs. This has put a strong emphasis on the importance of quality maintenance services used to care for systems. Service and maintenance are becoming essential for companies to sustain manufacturing productivity and customer satisfaction at the highest possible level.

INTRODUCTION

One of the most important tools in minimizing downtime, whether or not a conventional preventive maintenance program is possible, is called "preventive engineering." Although this would appear to be the application of common sense to equipment design maintenance engineering, it is a field which is often neglected. Too often maintenance engineers are so busy

DOI: 10.4018/978-1-5225-3244-6.ch001

handling emergency repairs or in other day-to-day activities that they find no opportunity to analyze the causes for breakdowns which keep them so fully occupied. While most engineers keep their eyes open to details such as better packings, longer-wearing bearings, and improved lubrication systems, true preventive engineering goes further than this and consists of actually setting aside a specific amount of technical manpower to analyze incidents of breakdown and determine where the real effort is needed; then through redesign, substitution, changes, and specifications, or other similar means, reducing the frequency of failure and the cost of repair (Linderly, Higgins & Mobley, 2002).

The high costs in maintaining today's complex and sophisticated equipment make it necessary to enhance modern maintenance management systems. Conventional condition-based maintenance (CBM) reduces the uncertainty of maintenance according to the needs indicated by the equipment condition. Intelligent predictive decision support system (IPDSS) for CBM supplements the conventional CBM approach by adding the capability of intelligent condition-based fault diagnosis and the power of predicting the trend of equipment deterioration. An IPDSS model, based on the recurrent neural network (RNN) approach, was developed and tested and run for the critical equipment of a power plant (Yam, Tse, Li & Tu, 2001).

These valuable results could be used as input to an integrated maintenance management system to pre-plan and pre-schedule maintenance work, to reduce inventory costs for spare parts, to cut down unplanned forced outage and to minimise the risk of catastrophic failure (Yam, Tse, Li & Tu, 2001). As soon as a bad condition is shown intelligently then react to it as soon as you see it.

In this research a move from the conventional CBM is needed to come up with intelligent CBM using fuzzy logic approach.

RESEARCH BACKGROUND

Condition-based maintenance (CBM) is a maintenance program that recommends maintenance decisions based on the information collected through condition monitoring. It consists of three main steps: data acquisition, data processing and maintenance decision-making. Diagnostics and prognostics are two important aspects of a CBM program. Research in the CBM area grows rapidly. Hundreds of papers in this area, including theory and practical applications, appear every year in academic journals, conference proceedings and technical reports (Shahanaghi, Babaei, Bakhsha & Fard, 2008).

Figure 1. Maintenance advice is made in accordance with equipment condition Yam, Tse, Li & Tu, 2001.

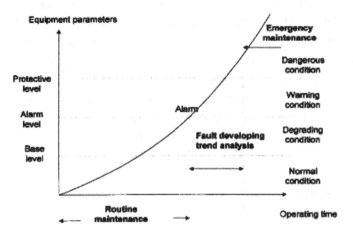

Condition Based Maintenance

Condition Based Maintenance (CBD) is a management philosophy that posits repair or replacement decisions based on equipment on the current or future condition of assets (Raheja, Llinas, Nagi & Romanowski, 2006); it recognizes that change in condition and/or performance of an asset is the main reason for executing maintenance (Horner, Elharam & Munns, 1997). CBM is a modern procedure which uses the condition of equipment to determine what, if any, testing and maintenance procedures should be performed (International Atomic Energy Agency, 2007). CBM is similar to preventive maintenance (PM) program which includes an extensive array of predictive maintenance (Pd.M.) procedures, so that necessarily means CBM is not Pd.M. but a Pd.M. is a subset of CBM.

$PdM + CBM = Holistic\ maintenance$ (This can have intelligence applied in it)

Figure 2. Three steps in a CBM program

As shown in Figure 3, the CBM approach has proactive and a predictive maintenance. All the problems and failures are analysed and solved systematically. Condition based is a holistic approach to maintenance. CBM only can reduce failures to a certain extent unless fused with intelligence in a way. On its own it can work with sensors and reduce further problems but is not much effective. With this current world of new technologies, CBM with intelligence will be a very good solution for reducing machine failures. The author selected fuzzy logic amongst other Artificial Intelligence (AI) approaches as explained in detail in the literature review.

Figure 3. CBM approach has a proactive and predictive maintenance approach.
Khobbacy & Murthy, 2007.

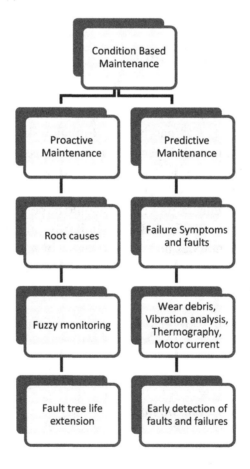

Are There Any Uncertainties/Fuzziness Experienced in CBM?

In all the condition monitoring techniques that are carried in most companies, it has dawned to the author that most monitoring techniques will be done sometimes when the machinery is malfunctioning or torn. Some areas of condition like thermography, when the experts come and do the process they will find some equipment damaged already, hence the application of intelligent real-time monitoring is necessary using fuzzy logic.

Did Any of the Gurus in the Literature Try to Tackle the Problem of Fuzziness/Ambiguity in CBM?

Large parts of our lives are now being monitored and analysed by computers and algorithms (Bournemouth University, 2013). Professor Gabrys is not so sure: "If someone tells you they can reliably predict really complex systems such as economies or financial markets one year ahead, do not believe them. Some things are predictable and some are not. The critical aspect in what we do is in knowing the difference between them (Bournemouth University, 2013)." Intelligence-based maintenance can eliminate or substantially reduce the risk of an accident. This requires the plant staff to know:

- The manufacturing process.
- The safety requirement of the process.
- The proper repair of equipment and components.
- How to identify and correct problem areas.

Three examples of fuzzy control are:

- The first is the control of a complex system.
- The second is a non-linear control system.
- The third is a medical diagnosis and treatment support system (Almardy, 1999).

The author will concentrate on the complex control of a plant to minimise plant failures. Maintenance must be intelligent (Netzel, 2011). Management, operations and maintenance must have open lines of communication to promote a culture of safety. Furthermore, educational/training programs need to be implemented for all plant personnel including all on-site contract working staff. If CBM is connected to fuzzy logic then operations will be closely monitored and breakdowns will lessen.

The productivity and output levels of construction plant and equipment depends in part upon a plant operator's maintenance proficiency; such that a higher degree of proficiency helps ensure that machinery is maintained in good operational order. In the absence of maintenance proficiency, the potential for machine breakdown (and hence lower productivity) is greater. Using data gathered from plant and equipment experts within the UK, plant operators' maintenance proficiency are modeled using a radial basis function (RBF) artificial neural network (ANN). Results indicate that the developed ANN model was able to classify proficiency at 89 per cent accuracy using 10 significant variables. These variables were: working nightshifts, new mechanical innovations, extreme weather conditions, planning skills, operator finger dexterity, years' experience with a plant item, working with managers with less knowledge of plant/equipment, operator training by apprenticeship, working under pressure of time and duration of training period. It is proffered that these variables may be used as a basis for categorizing plant operators in terms of maintenance proficiency and, that their potential for influencing operator training programmes need to be considered (Edwards, Yang, Cabahug & Love, 2005).

Indeed, there has been vast interest in the applications of AI in the maintenance area as witnessed by the large number of publications in the area (Khobbacy & Murthy, 2007). Dhaliwal (1986) is one of the early ones that argued for the appropriateness of using AI techniques for addressing the issues of operating and maintaining large and complex engineering systems. Kobbacy (1992) discusses the useful role of knowledge based systems in the enhancement of maintenance routines. Over the years the applications of AI in maintenance grew to cover very wide area of applications using a variety of AI techniques. This can be explained by the individual nature of each technique. For example, GAs and NNs have the advantage of being useful in optimising complex and nonlinear problems and overcome the limitations of the classic "black box" approaches, where attempt is made to identify the system by relating system outputs to inputs without understanding and modelling the underlying process. Hence the widespread applications in the scheduling area and also in fault diagnosis.

Fuzzy logic therefore allows the representation of information of uncertain nature. It provides a framework in which membership of a category is graded and hence quantifies such information for mathematical modelling (Khobbacy & Murthy, 2007).

Fuzzy logic has been used in various applications in the maintenance area to deal with uncertainty. Oke and Charles-Owaba (2006) apply an FL control model to Gantt charting preventive maintenance scheduling. Basim and Imad (2003) use a fuzzy multiple criteria decision making to select in advance the most informative (efficient) maintenance approach, i.e. strategies, policies or philosophies. Braglia, Frosolini and Montanari (2003) adopt FL to help an approach to allow analysts formulating efficiently assessment of possible causes of failure in mode, effects and criticality analysis.

The most popular maintenance approaches, i.e. strategies, policies, or philosophies, using a fuzzy multiple criteria decision making (MCDM) evaluation methodology has been assessed. It was concluded that using the fuzzy MCDM, it would be possible to select in advance, the most informative (efficient) maintenance approach. Consequently, this leads to less planned replacements, and failures would be reduced to approximately zero and higher utilization of component life can be achieved. Thus, the maintenance department could contribute more to the business objectives throughout participating effectively in adding value to the production activities Basim and Imad (2003). Sudiaros and Labib (2002) investigated fuzzy logic approach to an integrated maintenance/production scheduling algorithm. Jeffries, Lai, Plantenberg & Hull (2001) develop an efficient hybrid method for capturing machine information in a packaging plant using fuzzy logic, fuzzy condition monitoring, in order to reduce wastage and maintenance overheads. Examples of fuzzy logic hybrid applications include the use of a KBS for bridge damage diagnosis which aims at providing information about the impact of design factors on bridge deterioration with fuzzy logic used to handle uncertainties (Zhao & Chen, 2001). Sinha and Fieguth (2006), propose a neuro-fuzzy classifier that combines fuzzy logic and NNs for the classification of defects by extracting features in segmented buried pipe images.

Almardy used a fuzzy control system to apply current to a series of anodes to protect a long-buried pipeline. The goal was to maintain protection, but at the same time to minimize the power used to protect the pipeline. The disturbances were localized rain events which increased the soil conductivity in small regions along the pipeline. Fuzzy control was used as modeling the widely variable soil conditions along a pipeline would be too complex for a practical control system. The fuzzy model to control an experimental pipeline with three anodes consisted of 126 rules. The results from simulation trials and experimental data agreed well and the controller gave adequate performance in maintaining protection. The controller gain, which determines the control

Figure 4. Fuzzy logic process in maintenance
Dash, Rengaswamy & Venkatasubramanian, 2003.

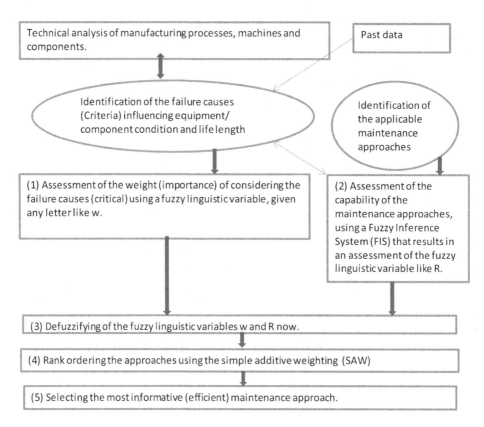

system stability, was tuned by adjusting the output membership functions (Almardy, 1999). This is a type of CBM approach to eliminate failures and breakdowns of the leakages to the pipeline. Fuzzy logic is doing online condition monitoring of the pipeline.

Other applications for fuzzy logic in fault diagnosis include fault diagnosis of railway wheels (Skarlatos, Karakasis & Trochidis, 2004), thrusters for an open- frame underwater vehicle (Omerdic & Roberts, 2004), chemical processes (Dash, Rengaswamy & Venkatasubramanian, 2003) and rolling element.

It has dawned to the researcher that fuzzy logic has been proven to be a viable alternative amongst other algorithms in reducing machinery failure from the already published journals and articles. Most authors used fuzzy logic in pure sciences like in physics, chemistry, computer science and biology. Using fuzzy logic in engineering is a really promising tool as seen for the work done in pipelines and railways in trying to do CBM. Fuzzy logic removes the fuzziness of a complex problem to simpler terms and solves faster.

Fuzzy Logic Systems

There are many misconceptions about fuzzy logic. To begin with, fuzzy logic is not fuzzy (Zadeh, 2008). In large measure, fuzzy logic is precise (Mathworks, 1994). Another source of confusion is the duality of meaning of fuzzy logic. In a narrow sense, fuzzy logic is a logical system. But in much broader sense which is in dominant use today, fuzzy logic, or FL for short, is much more than a logical system (Zadeh, 2012). It is one of components that make up an AI. The other ones are; expert systems, neural networks and genetic algorithm. Maintenance is fuzzy in nature in the sense that a failure can occur or not or vice versa.

The concept of fuzzy logic was introduced by Lotfi Zadeh, as a formal methodology to represent heuristic knowledge. Fuzzy logic controller systems are a practical alternative for the development of a wide variety of control application, being able to control nonlinear systems using heuristic information supplied intuitively by the programmer. Fuzzy controllers have robustness and low cost inherent characteristics. They can be implemented in hardware in several ways such as microprocessors or microcontrollers, but there are digital signal processor based implementations as well.

Fuzzy Logic is a form of logical reasoning that can be incorporated into automation systems typically human reasoning schemes. One of the main features of fuzzy logic is its ability to operate with vague or ambiguous concepts typical of qualitative reasoning, based on a mathematical support quantitative conclusion can be drawn from a set of observations and qualitative rules. The most important specifications of fuzzy logic control method are their fuzzy logical ability in the quality perception of system dynamics and the application of these quality ideas simultaneously for control system. A simple block diagram of a fuzzy logic system is shown in Figure 5.

Why Fuzzy Logic?

We definitely need fuzzy logic in CBM because of its nature from the beginning. One can set up a fuzzy system for the same purpose he/she set up any other computing system—to map inputs to outputs. Basically, it consists of three

Figure 5. Fuzzy logic flow process

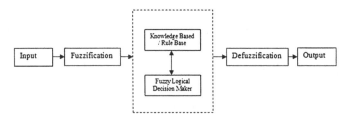

stages: fuzzification, rule evaluation, and defuzzification. Fuzzification is a process that combines actual values (e.g., barometric pressure) with stored membership-function data to produce fuzzy input values. Rule evaluation, or fuzzy inferencing, is a way of producing numeric responses from linguistic rules based on system input values. In the last stage—defuzzification—a fuzzy system combines all its outputs and obtains a representative number. To see if this number solves the original problem and gives one an accurate answer in all cases, Fred Watkins, president of HyperLogic, a firm that produces fuzzy-logic development tools, says it's necessary to come up with a performance measure (theoretically, an ideal correct response). One can then run the engine in a variety of contexts. If the number doesn't turn out to be a good solution, one can tune the system parameters until he/she reach a satisfactory conclusion. Even as the rules of a fuzzy engine become more complex, says Watkins, the general concepts remain the same.

According to Emdad Khan, manager of fuzzy and neural networks for the Embedded Systems Division of National Semiconductor, you can construct a PC-based fuzzy-logic system using software alone. However, general-purpose or dedicated microprocessors are available for more complicated applications. Natural language, which is used by ordinary people on a daily basis, has been shaped by thousands of years of human history to be convenient and efficient. Uncertainty processing is a fact of life. It is an important part of daily routines. Various theories have been proposed. Here is a list of general observations about fuzzy logic: (Zahra, 2014)

Fuzzy logic is conceptually easy to understand. The mathematical concepts behind fuzzy reasoning are very simple. Fuzzy logic is a more intuitive approach without the far-reaching complexity.

1. *Fuzzy logic is flexible.* With any given system, it is easy to layer on more functionality without starting again from scratch.

Table 1. Selection of fuzzy logic artificial intelligence tool

Intelligency Tool	Response Time	Scalability	Flexibility	Ease of Use	Embeddability	Processing Overhead	Expert Dependence	Tolerance for Dirty Data	Implementation Speed	Tolerance for Complexity	Accuracy
Genetic algorithm	H	L	L	L	H	H	L	-	H	H	H
Neural networks	H	M	L	L	M	L	L	H	M	H	H
Fuzzy logic systems	H	L	M	H	M	H	H	H	H	M	M
Rule-based systems	X	L	M	H	H	M	H	L	H	M	H
Case based reasoning	L	H	L	M	M	H	L	H	M	H	H

Mushiri, 2012.

Legend: H = High; L =Low; M = Medium; X = Deteriorates as the number of active rules grows

2. *Fuzzy logic is tolerant of imprecise data.* Everything is imprecise if you look closely enough, but more than that, most things are imprecise even on careful inspection. Fuzzy reasoning builds this understanding into the process rather than taking it onto the end (Mushiri, Manjengwa and Mbohwa, 2014).

3. *Fuzzy logic can model nonlinear functions of arbitrary complexity.* You can create a fuzzy system to match any set of input-output data. This process is made particularly easy by adaptive techniques like Adaptive Neuro-Fuzzy Inference Systems (ANFIS).

4. *Fuzzy logic can be built on top of the experience of experts.* In direct contrast to neural networks, which take training data and generate opaque, impenetrable models, fuzzy logic lets you rely on the experience of people who already understand your system.

5. *Fuzzy logic can be blended with conventional control techniques.* Fuzzy systems don't necessarily replace conventional control methods. In many cases fuzzy systems augment them and simplify their implementation.

6. *Fuzzy logic is based on natural language.* The basis for fuzzy logic is the basis for human communication. This observation underpins many of the other statements about fuzzy logic. Because fuzzy logic is built on the structures of qualitative description used in everyday language, fuzzy logic is easy to use (Zahra, 2014).

Natural language, which is used by ordinary people on a daily basis, has been shaped by thousands of years of human history to be convenient and efficient. Sentences written in ordinary language represent a triumph of efficient communication.

Therefore, fuzzy logic was selected as the best way to reduce machinery failure and synergized with CBM.

REFERENCES

Almardy, M. (1999). *Design of fuzzy logic controller for the cathodic protection of underground pipelines.* University of Guelph.

Basim, A., & Imad, A. (2003). Selecting the most efficient maintenance approach using fuzzy multiple criteria decision making. *International Journal of Production Economics, 84*(1), 85–100. doi:10.1016/S0925-5273(02)00380-8

Bournemouth University. (2013, March 13). Predictive analysis: New Generation of Computational Intelligence Systems. *Science Daily.*

Braglia, M., Frosolini, M., & Montanari, R. (2003). Fuzzy criticality assessment model for failure modes and effects analysis. *International Journal of Quality & Reliability Management, 20*(4), 503–524. doi:10.1108/02656710310468687

Dash, S., Rengaswamy, R., & Venkatasubramanian, V. (2003). Fuzzy logic based trend classification for fault diagnosis of chemical processes. *Computers & Chemical Engineering, 27*(3), 347–362. doi:10.1016/S0098-1354(02)00214-4

Dhaliwal, D. S. (1986). The use of AI in maintaining and operating complex engineering systems. In *Expert systems and optimisation in process control* (pp. 28-33). Gower Technical Press.

Edwards, D. J., Yang, J., Cabahug, R., & Love, P. E. D. (2005). Intelligence and maintenance proficiency: An examination of plant operators. *Construction Innovation: Information, Process, Management, 5*(4), 243–254. doi:10.1108/14714170510815285

Horner, R. M. W., Elharam, M. A., & Munns, A. K. (1997). Building maintenance strategy: A new management approach. *Journal of Quality in Maintenance Engineering, 3*(4), 273–280. doi:10.1108/13552519710176881

International Atomic Energy Agency. (2007). *Implementation Strategies and Tools for Condition Based Maintenance at Nuclear Power Plants.* Vienna: International Atomic Energy Agency.

Jeffries, M., Lai, E., Plantenberg, D. H., & Hull, J. B. (2001). A fuzzy approach to the condition monitoring of a packaging plant. *Journal of Materials Processing Technology, 109*(1-2), 83–89. doi:10.1016/S0924-0136(00)00779-2

Khobbacy, A. H. K., & Murthy, P. D. N. (2007). *Complex system maintenance handbook.* Springer London.

Kobbacy. (1992). The use of knowledge-based systems in evaluation and enhancement of maintenance routines. *International Journal of Production Economics,* 243-248.

Linderly, R., Higgins, R., & Mobley, K. (2002). *Maintenance Engineering Handbook* (6th ed.). McGraw Hill.

Mathworks. (1994, December 20). What Is Fuzzy Logic? Retrieved March 3, 2015, from http://www.mathworks.com/help/fuzzy/what-is-fuzzy-logic.html

Mushiri. (2012). *A study into the role of intelligent systems in coming up with a condition based maintenance framework at delta beverages, sparkling branch, Harare, Zimbabwe.* Harare: University of Zimbabwe.

Mushiri, Manjengwa, & Mbohwa. (2014). Advanced Fuzzy Control In Industrial Wastewater Treatment (pH and temperature control). In *World Congress of Engineering, WCE 2014* (pp. 53-58). London: Newswood Limited, Organization: International Association of Engineers (IAENG).

Netzel. (2011, April 11). *Maintenance technology: The source of reliable solutions.* Retrieved March 5, 2013, from http://www.mt-online.com/april2011/all-maintenance-must-be-intelligence-based

Oke & Charles-Owaba. (2006). Application of fuzzy logic control model to Gantt charting preventive scheduling. *International Journal of Quality & Reliability Management*, 441–459.

Omerdic, E., & Roberts, G. (2004). Thruster fault diagnosis and accommodation for open frame underwater vehicles. *Control Engineering Practice*, *12*(12), 1575–1598. doi:10.1016/j.conengprac.2003.12.014

Raheja, D., Llinas, J., Nagi, R., & Romanowski, C. (2006). Data fusion / data mining based on architecture for condition based maintenance. *International Journal of Production Research*, 717–728.

Shahanaghi, K., Babaei, H., Bakhsha, A., & Fard, N. S. (2008). A new Condition Based Maintenance model with random improvements on the system after maintenance actions: Optimizing by monte carlo simulation. *World Journal of Modelling and Simulation*, *4*(3), 230–236.

Sinha, S. K., & Fieguth, P. W. (2006). Neurofuzzy network for the classification of buried pipe defects. *Automation in Construction*, *15*(1), 73–83. doi:10.1016/j.autcon.2005.02.005

Skarlatos, D., Karakasis, K., & Trochidis, A. (2004). Railway wheel fault diagnosis using fuzzy logic method. *Applied Acoustics*, *65*(10), 951–966. doi:10.1016/j.apacoust.2004.04.003

Sudiaros, A., & Labib, A. W. (2002). A fuzzy logic approach to an integrated maintenance/production scheduling algorithm. *International Journal of Production Research*, *40*(13), 3121–3138. doi:10.1080/00207540210146143

Yam, R. C. M., Tse, P. W., Li, L., & Tu, P. (2001). Intelligent predictive decision support system for condition-based maintenance. *International Journal of Advanced Manufacturing Technology, 17*(1), 383–391. doi:10.1007/s001700170173

Zadeh. (2008). Is there a need for fuzzy logic? *Information Sciences, 178*, 2751-2779.

Zadeh. (2012, February 1). *Fuzzy logic*. Retrieved February 4, 2015, from Fuzzy logic: http://www.scholarpedia.org/article/Fuzzy_logic

Zahra. (2014, February 6). *Fuzzy Logic Control Tutorial*. Retrieved March 10, 2015, from https://sites.google.com/site/controlandelectronics/fuzzy-logic-control-tutorial

Zhao, Z., & Chen, C. (2001). Concrete bridge deterioration diagnosis using fuzzy inference system. *Advances in Engineering Software, 32*(4), 317–325. doi:10.1016/S0965-9978(00)00089-2

KEY TERMS AND DEFINITIONS

AI: Artificial intelligence.
ANN: Artificial neural network.
CBM: Condition-based maintenance.
IPDSS: Intelligent predictive decision support system.
ISO: International Standards Organisation.
PdM: Predictive maintenance.
RBF: Radial basis function.

Chapter 2
The Case for Condition–Based Monitoring

ABSTRACT

Maintenance is the management, implementation, and control of all activities that are meant to restore an asset to its ideal level of performance and availability with a view to meet the company objectives and competitive advantages. Maintenance has a greatest contribution to profits within an organisation. The profitable and competitive environment of today demands full productivity in industries with minimum capital investments. This involves coming up with strategies to maximise the up time and machine reliability, extending the life of both the plant and the equipment through cost-effective maintenance.

INTRODUCTION

Maintenance Before World War Two

During pre-World War II, people assumed of maintenance as an additional cost to the plant without increasing the value of finished goods. Therefore, maintenance was restricted to fixing the machine when after break down because it was the cheapest substitute. In other words, people used to manufacture and delay for the equipment to develop some failure then start maintenance. Worker could communicate with each other before changing shifts so that the incoming shift will be aware of all possible harms and dangerous areas (Napp, 2009).

DOI: 10.4018/978-1-5225-3244-6.ch002

Maintenance During World War Two

Engineers and engineering students who were recruited to join the military were not present for work. As a result, the United States Office of Education arranged a number of maintenance programs. These programs include the Engineering, Science and Management Defence Training (ESMDT) or Engineering, Science and Management War Training (ESMWT) and Engineering Defence Training. All university conducted all these courses with the Office of Education paying for instruction, laboratory equipment, and maintenance (Napp, 2009).

Maintenance After World War Two

During this period, maintenance, automation and breakdown of machinery was a major concern. Reliability Centred Maintenance (RCM) was introduced in 1964 and Total Productive Maintenance (TPM) in the 1971s by Japan (Moballeghi M, 2013). The tendency of maintenance management was increased in plant machinery using the concept of *kaizen, a* Japanese term representing continuous improvement in production or manufacturing. In today's industries, increased awareness in issues such as safety, product quality, maintenance and environment are considered to be major functions that can contribute to the success of the industry. (Napp, 2009). The history of maintenance is well described in Figure 1.

Condition monitoring (CM) is the process of monitoring a parameter of condition in machinery, such that a significant change is indicative of a developing failure, i.e. it uses statistics (Okah-Avae, 1981). It is a major component of predictive maintenance. The use of conditional monitoring allows maintenance to be scheduled, or other actions to be taken to avoid the consequences of failure, before the failure occurs. Nevertheless, a deviation from a reference value (e.g. temperature or vibration behavior) must occur to identify impeding damages (Jardine, Daming and Dragan, 2006). Predictive Maintenance does not predict failure. Machines with defects are more at risk of failure than defect free machines. Once a defect has been identified, the failure process has already commenced and CM systems can only measure the deterioration of the condition. Intervention in the early stages of deterioration is usually much more cost effective than allowing the machinery to fail. Condition monitoring has a unique benefit in that the actual load, and subsequent heat

Figure 1. Maintenance history

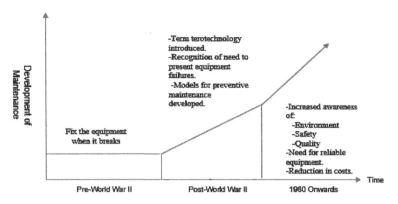

dissipation that represents normal service can be seen and conditions that would shorten normal lifespan can be addressed before repeated failures occur. Serviceable machinery includes rotating equipment and stationary plant such as boilers and heat exchangers (Liu, Jie, Wang and Golnaraghi, 2008). Condition Based Maintenance on the other hand is to monitor and assess the condition or health of a machinery unit while it is running and stop it for maintenance only (Okah-Avae, 1981). On-load monitoring is done without interruption of the operating unit and off - load monitoring, which would require the unit to be shut down or at least removed from its prime duties. Figure 9 shows the available prognostic methods and its groups. Condition-based maintenance is a special form of preventive maintenance that is based on performance parameter monitoring and subsequent actions (Nilsson et al, 2007). According to this definition, condition monitoring is partly an optimised component of preventive maintenance.

NASA (2008) outlines the condition monitoring processes as:

- Trend analysis
- Pattern recognition
- Data comparison
- Tests against limits and ranges
- Correlation of multiple technology
- Statistical process analysis

There are different types of sensors which uses different principles to collect data. These sensors produce different types of signals as input to controllers.

Figure 2. Condition monitoring methods and their groups
Okah-Avae, 1981.

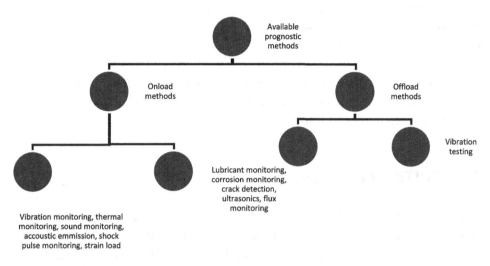

Currently, driven by the demand to reduce maintenance costs, shorten repair time, and maintain high availability of equipment, maintenance strategies have progressed from breakdown maintenance (fail and fix) to preventive maintenance (Lee, Ni, Djurdjanovic, Qiu and Liao, 2006), then to condition-based maintenance (CBM), and lately toward a prospect of intelligent predictive maintenance (predict and prevent) (Heng, Zhang, Tan and Matthew, 2008), (Tao, Chen, Chan and Wang, 2013). In actual fact the breakdown maintenance and preventive maintenance are labor intensive and also expensive to keep on doing them. For rotating machinery, it will be very difficult however to apply PM and breakdown maintenance hence the need to come up with intelligent CBM. Generally, this section of literature highlights the other work and case studies that have been carried out in doing maintenance to move away from the general maintenance that is found now to be expensive and time consuming.

The idea underlying conditional maintenance is that a component is not replaced unless it is showing signs of ageing or wear likely to impair its performance (Blanchard and Fabrycky, 1990). The decision is based on physical measurements which are usually:

- Noise and vibration
- Temperatures (thermal monitoring)

- Analysis of oil debris
- Corrosion monitoring
- Acoustic emission monitoring
- Motor and transformer current analysis

The more specialized methods which tend to be associated with particular plant or industry include ultrasonics, shock pulse monitoring, crack detection and some more advancement (Okah-Avae, 1981).

THE CRITICALITY INDEX

The Criticality Index is often used to determine the degree on condition monitoring on a given machine taking into account the machines purpose, redundancy (i.e. if the machine fails, is there a standby machine which can take over), cost of repair, downtime impacts, health, safety and environment issues and a number of other key factors. The criticality index puts all machines into one of three categories:

1. **Critical Machinery:** Machines that are vital to the plant or process and without which the plant or process cannot function. With critical machinery being at the heart of the process it is seen to require full on-line condition monitoring to continually record as much data from the machine as possible regardless of cost and is often specified by the plant insurance. Measurements such as loads, pressures, temperatures, casing vibration and displacement, shaft axial and radial displacement, speed and differential expansion are taken where possible. These values are often fed back into a machinery management software package which is capable of trending the historical data and providing the operators with information such as performance data and even predict faults and provide diagnosis of failures before they happen.
2. **Essential Machinery:** Units that are a key part of the process, but if there is a failure, the process still continues. Redundant units (if available) fall into this realm. Testing and control of these units is also essential to maintain alternative plans should Critical Machinery fail.
3. **General Purpose or Balance of Plant Machines:** These are the machines that make up the remainder of the plant and normally monitored using a handheld data collector as mentioned previously to periodically create a picture of the health of the machine.

TYPES OF FAILURE CAUSED BY NOT MAINTAINING MACHINERY

Failures may be classified in terms of criticality in any one of four categories, depending on the defined failure effects as follows.

- **Minor Failure:** Any failure that does not degrade the overall performance and effectiveness of the system beyond acceptable limits.
- **Major Failure:** Any failure that will degrade the system performance and effectiveness beyond acceptable limits but van be controlled.
- **Critical Failure:** Any failure that will degrade the system beyond acceptable limits and could create a safety hazard if immediate corrective action is not taken.
- **Catastrophic Failure:** Any failure that could result in significant system damage, such as to preclude functional accomplishment, and could cause deaths and personnel injuries (Blanchard, 1990).

TYPES OF CONDITION MONITORING TECHNIQUES

In line with the research area of interest, condition monitoring will be looked at since it is a branch of CBM.

Condition Monitoring Techniques

Noise and Vibration

All machines vibrate, and when they are in good condition their frequency spectrum has a characteristic form, any departure from this form indicates that something is wrong- fatigue, or wear, or ageing of something of some component. Parameters that are useful include amplitude, frequency and phase angle (Okah-Avae, 1981).

Amplitude

This gives an indication of the stress under which a piece of rotating machinery is working, in particular, it can give a measure of the eccentricity (out-of –roundness) of a rotor (Okah-Avae, 1981).

Frequency

Through the frequency spectrum, you can detect a fault in rotating machinery. Vibrations fall into two main classes:

- **Synchronous:** Frequencies are in multiples or sub-multiples of the frequency of rotation that is they are harmonics or sub- harmonics of that frequency.
- **Asynchronous:** These are not related to the rotation frequency, they can be the natural frequencies of various parts of the system, which can be identified (Okah-Avae, 1981).

Phase Angle

It locates the high point in a rotor that is not perfectly circular, and thus gauges its out-of-balance characteristics. Machines must be continuously monitored to denote their state. The following machines can do these:

1. **Clearance Recorder:** Recording the actual movements of the shaft which generate the vibrations.
2. **Speed Recorder:** Mounted externally to a machine and it gives a strong signal at medium frequencies, depending on the temperature and the general environment.
3. **Accelerometer:** Also installed to a machine and it gives a strong signal at high frequencies (Okah-Avae, 1981).

Temperature

Temperature recording is a relatively simple matter at the industrial level. Change of temperature in rotating machinery is often a sign of deterioration, and is therefore something to which close attention should be given. This is currently practiced in thermography.

Tribology

An examination of the particles suspended in the oil can give very valuable information. The amount of suspended material is an indicator of the state of deterioration of the machine; the composition can identify the source of the wear and thus the component that is failing. The necessary analysis can be done in the laboratory with the electron microscope.

Summary of Condition Monitoring Techniques

Basically, the monitoring techniques that were listed in Table 1 are almost the major ones currently used.

Objectives of Condition Based Maintenance

In the use of CBM there are quite a number of objectives need to be achieved as a further way of maintenance from Preventive Maintenance. Some of them are as follows.

- Giving good indication of when a machine is running smoothly and efficiently
- To intervene shortly before failure occurs
- To do maintenance only when needed
- To reduce number of failure and also number of shutdowns
- Elimination of secondary damage
- To reduce maintenance cost
- Increase life of equipment
- Improvement in safety, product quality and customer relations
- Reduction in inventory cost/effective inventory control.

Benefits of Condition Based Maintenance (CBM)

Inasmuch as CBM has been in place since its introduction, a lot of advantages were noticed and it's one of the modern ways of accurate and precise maintenance.

Table 1. Summary of condition monitoring techniques

Type	Method
Visual	Eye
Temperature	Thermometer, thermocouple
Lubricant monitoring	Filtering, spectroscopy
Vibration	Signal frequency analysis
Crack	Di-penetrant analysis, radiography
Corrosion monitoring	Eye, corrosometer

- Reduces the likelihood of maintenance induced failures by increasing maintenance intervals
- Lowers inventory levels since parts can be ordered when needed
- Allows scheduling of maintenance to consider production needs. Thus, reducing lost production due to maintenance downtimes
- Improved system availability
- Improved plant operation and safety
- Improved maintenance
- Improved product quality

Furthermore, condition-based maintenance implies that maintenance activities be scheduled in a dynamic way, since the execution times of certain activities will be continually updated as condition information becomes available. Given the complexity of the processes underlying mechanical and structural degradation and the ambiguous and uncertain character of the experimental data available, one may have to resort to empirical models based on collected evidence, some of which may very well be of qualitative, linguistic nature. In this direction, soft computing techniques, such as neural networks and fuzzy logic systems (inferential systems based on the mathematics of fuzzy sets), represent powerful tools for their capability of representing highly non-linear relations, of self-learning from data and of handling qualitative information (Zio, 2007).

Condition Monitoring Architecture

Online Condition Monitoring Architecture

Online condition monitoring has been implemented using various solutions by various vendors. The bulk of such systems is developed in developed countries and is not widely used in the electricity industry in the developing nations, condemning the industry to a competitive dis advantage. Figure 3 shows one such condition monitoring system implemented by Bently Nevada on a hydro-generator. Such systems have a major drawback which is going to be addressed by this book. The system is based on step by step logical computer automation and does not make use of the full capabilities of intelligent systems. The data generated by this system should be interpreted by an engineer for it to be used for decision making. The system to be designed by way of this dissertation behaves like an Expert Reliability Engineer and hence can make decisions, provide advice, answers to questions and gives explanations.

Figure 3. Condition monitoring of a turbo-gen-set
Bentley Nevada, 2013.

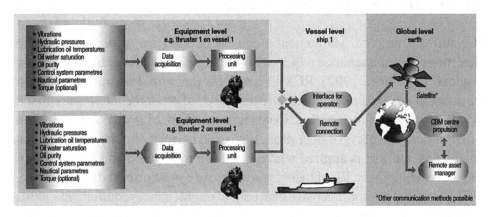

Turbo Alternator Condition Monitoring System

Several signals can be detected by having online condition monitoring for Hydro- Alternator sets and the necessary sensors. The conditions include mechanical unbalance, electrical unbalance, shear pin failure, misalignment, cavitation, turbine vibration, insufficient bearing lubrication, stator insulation deterioration, uneven air gap and others. Different types of sensors are used to detect the conditions using different technologies which are beyond the scope of this literature.

Intelligent Condition Based Maintenance

Sethiya (2013) defines an intelligent CBM as a CBM system capable of understanding and making decisions without human intervention. Algorithms, fuzzy logic and neural networking analyse trends and produce decisions on the likelihood of failure of the monitored plant. The artificial intelligence can provide proxy data as a substitute for failed sensor whilst the sensor is being repaired.

CBM AS A BRANCH OF RELIABILITY
CENTRED MAINTENANCE (RCM)

Reliability Centered Maintenance can be defined as "a process used to determine the maintenance requirements of any physical asset in its operating context" (Moubray, 1997). RCM is a process that will determine what must be done, in terms of maintenance, to any piece of equipment to ensure that it continues to fulfill its function. Consideration of RCM is given to safety and environmental aspects of operation. Improved performance by ensuring that maintenance effort is applied where it will do most good and longer life of expensive equipment. RCM focuses on the use of condition based maintenance. RCM develops a comprehensive data base of maintenance requirements, skills required and stocks that should be held. Over the past fifteen years, maintenance has changed, perhaps more so than any other management discipline. The changes are due to a huge increase in the number and variety of physical assets (plant, equipment and buildings) which must be maintained throughout the world, much more complex designs, new maintenance techniques and changing views on maintenance organisation and responsibilities. Maintenance is also responding to changing expectations. These include a rapidly growing awareness of the extent to which equipment failure affects safety and the environment, a growing awareness of the connection between maintenance and product quality, and increasing pressure to achieve high plant availability and to contain costs. The changes are testing attitudes and skills in all branches of industry to the limit. Maintenance people have to adopt completely new ways of thinking acting, as engineers and as managers. At the same time the limitations of maintenance systems are becoming increasingly apparent, no matter how much they are computerised.

In the face of this avalanche of change, managers everywhere are looking for a new approach to maintenance. They want to avoid the false starts and dead ends which always accompany major upheavals. Instead they seek a strategic framework which synthesizes the new developments into a coherent pattern, so that they can evaluate them sensibly and apply those likely to be of most value to them and their companies. If it is applied correctly, RCM transforms the relationships between the undertakings which use it, their existing physical assets and the people who operate and maintain those assets. It also enables new assets to be put into effective service with great speed, confidence and precision. RCM is rapidly becoming a cornerstone of the Third Generation Maintenance, but this generation can only be viewed in perspective in the light of the First and Second Generations.

Reliability-centered maintenance (RCM) is the optimum mix of reactive, time or interval-based, condition-based, and proactive maintenance practices. These principal maintenance strategies, rather than being applied independently, are integrated to take advantage of their respective strengths in order to maximize facility and equipment reliability while minimizing life-cycle costs. Reliability- centered maintenance (RCM), and many other innovative approaches to maintenance problems all aim at enhancing the effectiveness of machines to ultimately improve productivity (Shayeri, 2007).

Reactive maintenance is usually only implemented following an unforeseen event leading to a partial or total failure of the system. Preventive maintenance (PM) is initiated according to a predetermined time-schedule in order to try to avoid the occurrence of failure. Predictive maintenance (Pd.M.) is launched as a result of behaviour of the equipment/ machinery before total failure, whereas proactive maintenance may require redesigning and/or modification of the adopted maintenance-procedure where necessary (Ugechi, Ogbonnaya, Lilly, Ogaji and Probert, 2009).

This chapter aims at the establishment and analysis of the necessary literature in the development of an (Expert Reliability Centered Maintenance System) ERCM. An assessment of the works already done on the subject under discussion shall be done with particular reference to the ideas presented in publications and journals by different researchers in the field of RCM and Artificial intelligence. The major artificial intelligence systems like the Expert systems, fuzzy logic, neural network and genetic algorithm were discussed in general to support the choice of Expert systems in the implementation of an Intelligent RCM System. The researcher reviewed the literature about the current trends on Expert systems and RCM processes.

Expert systems are based on expert knowledge about the subject at hand and hence it is imperative that literature on RCM be discussed and be

Figure 4. CBM as a branch of RCM
Adopted from (Ugechi, Ogbonnaya, Lilly, Ogaji and Probert, 2009).

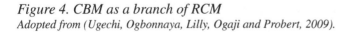

Figure 5. Components of RCM program
Ugechi, Ogbonnaya, Lilly, Ogaji and Probert, 2009.

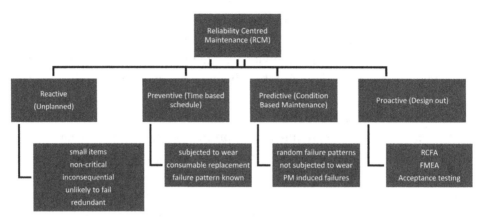

reviewed extensively as it shall be used in knowledge acquisition, analysis and extraction tasks. Giarratano and Riley (1998) state the main objective of the knowledge acquisition, analysis and extraction tasks as to produce and verify the knowledge required by the system. The knowledge definition is also critical as it defines the knowledge requirements of an expert system.

This chapter has presented what constitute an Expert System. RCM and intelligent systems as viewed by other authors has been presented. Works on maintenance systems based on Artificial Intelligence has been undertaken though more research is required to cover the new methodologies like RCM especially as applied to the Hydro-power generation.

Table 2 shows the Stan Nowlan and Harold Heap curves that apply to almost all the machinery around the globe.

The "condition" being measured can take a variety of forms. Any condition that shows a change, as the health of the spare deteriorates, can be used. It must however; give enough of a warning between "P" and "F" to allow actions to take place otherwise nothing will be gained by having a warning. Equipment condition being measured could include:

- Reducing pressure supplied by a pump indicating impeller wear, slip ring wear, etc.
- increased surface temperature on the outside of an insulating surface indicating insulation deteriorating
- Increased vibration from rotating equipment; these vibrations must be analysed further to identify possible causes.

Table 2. Six classic failure patterns

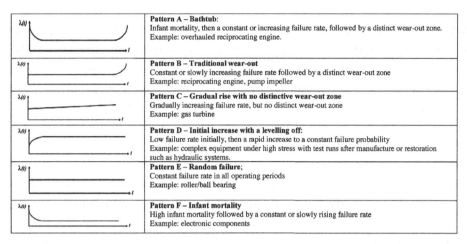

	Pattern A – Bathtub: Infant mortality, then a constant or increasing failure rate, followed by a distinct wear-out zone. Example: overhauled reciprocating engine.
	Pattern B – Traditional wear-out Constant or slowly increasing failure rate followed by a distinct wear-out zone Example: reciprocating engine, pump impeller
	Pattern C – Gradual rise with no distinctive wear-out zone Gradually increasing failure rate, but no distinct wear-out zone Example: gas turbine
	Pattern D – Initial increase with a levelling off: Low failure rate initially, then a rapid increase to a constant failure probability Example: complex equipment under high stress with test runs after manufacture or restoration such as hydraulic systems.
	Pattern E – Random failure; Constant failure rate in all operating periods Example: roller/ball bearing
	Pattern F – Infant mortality High infant mortality followed by a constant or slowly rising failure rate Example: electronic components

Nowlan and Heap, 1978.

Figure 6. The P-F Curve showing inspection intervals
Sondalini, 2014.

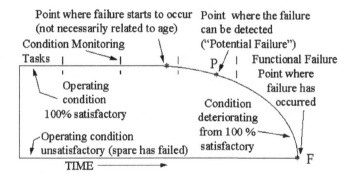

Having identified point P then two actions can take place:

- **To Prevent the Functional Failure:** Depending on the nature of the failure mechanism, it is sometimes possible to intervene to repair the existing component before it fails completely.
- **To Avoid the Consequences of the Failure:** In most cases, detecting a potential failure does not actually prevent the spare from failing, but still makes it possible to avoid or reduce the consequences of the failure. For example the necessary spares, personnel and equipment could be made available, or the effected part could be changed out of production time before it actually fails.

Background of RCM and Intelligent Systems

A lot of research has been done around reliability centred maintenance with remarkable literature being published in the early 80's by the likes of Moubray, Nowlan and Heap. RCM in itself is a body of knowledge for optimum maintenance of facilities equipment. Drastic equipment failure has happened in the past and has resulted in loss of life, damage to the environment, service stoppage and big economic losses. Some of the notable equipment failures include the Deep Water Horizon incidence, Chernobyl and the Challenger disaster among hundreds others.

Most of the research and works done by the RCM pioneers has been largely around RCM as a body of knowledge on asset management. The works has generally published the methodology and processes without necessarily paying attention on ways of applying intelligent systems in carrying out the RCM procedures and decision making. This is probably because RCM is

Table 3. Some reliability parameters

Parameters	Expressions
Mean Time to Failure	$MTTF_s = \dfrac{1}{\lambda_s}$ (1)
Mean Time to Repair	$MTTR_s = \dfrac{1}{m_s} = t_s$ (2)
Mean Time Between Failures	$MTBF_s = MTTR_s + MTTF_s$ (3)
Availability	$A_s = \dfrac{m_s}{(m_s + \lambda_s)} + \dfrac{\lambda_s}{(m_s + \lambda_s)} e^{-(m_s + \lambda_s)t}$ (4)
Reliability	$R_s = e^{-(\lambda_s t)}$ (5)
Expected number of failures	$W_s = \dfrac{\lambda_s m_s t}{(m_s + \lambda_s)} + {\lambda_s^2}\Big/{(m_s + \lambda_s)^2}\left[1 - e^{-(m_s + \lambda_s)t}\right]$ (6)

Source: (Kumar, Sharma & Kumar, 2007).
Key: Where Failure rate = Lambda = λ = f/hr (per hour)

relatively a new asset management methodology in the industry and is yet to be explored further.

Much of the literature available on RCM refers to the aeronautical and airline industry. This may be partly because the pioneers and the gurus of this methodology belongs to the industry and little effort has been put by those in the electricity generation industry to research around the subject.

Martin (2011) notes that RCM started in the airline industry because traditional methods based on preventing asset deterioration rather than minimising failure consequences were too costly and not effective and hence exposed the aerospace industry to viability challenges. This chapter is going to review the works already done around RCM and Intelligent systems separately. These two distinct fields are going to be married together to formulate an RCM based on intelligent systems in the later chapters. RCM computer systems already exist in the market; these include such advanced systems like the NAVAIR's IRCMS 6.3 which provides users with an aid to perform Reliability Centred Maintenance (RCM) analysis. This Microsoft Access-based application provides multiple user access to the analysis package that aids in describing the system being analysed, failure modes and effects analysis, and cost calculations for a variety of maintenance options (Gernand 2008). Gernand et al (2008) noted the weaknesses to of IRCM 6.3 as being labour intensive plus the difficulty with data inputting as all the data is entered manually and there is no automatic information gathering as by the online condition monitoring system. These are some of the weaknesses which call for continued research, growth and development of RCM systems.

The aim of the RCM system is to integrate each of the maintenance concepts into a single optimum maintenance system. The term Artificial intelligence was coined by John McCarthy at the Massachusetts Institute of Technology in 1956. Artificial Intelligence is the study of man-made computational devices and systems which can be made to act in a manner which we would be inclined to call intelligent. It is the branch of computer science concerned with making computers behave like humans ('Artificial Intelligence,' n.d.). Krishnakumar (2000) expanded the definition of intelligent system as one that emulates some aspects of intelligence exhibited by nature. These include:

1. Learning
2. Adaptability
3. Robustness across problem domains
4. Improving efficiency (over time and/or space)

5. Information compression (data to knowledge)
6. Extrapolated reasoning

A system is intelligent if it accomplishes feats that, when carried out by humans, require a substantial amount of intelligence. An intelligent agent is an intelligent system which perceives its environment by sensors and which uses that information to act upon the environment (Truemper, 2004).

Why Expert Systems?

Four major branches of Artificial Intelligence are to be discussed and compared as shown below:

1. Fuzzy logic
2. Expert systems
3. Genetic Algorithms
4. Artificial Neural Networks

RCM is a highly specialized field and depends much on expert knowledge and hence the selection of Rule based expert systems. When making RCM decisions, fuzzy reasoning is common as information is often incomplete and uncertainty high. In such cases, fuzzy expert systems become paramount in the development of an ERCM. Expert systems are best suited for those dealing with expert heuristics for solving problems (Chakraborty 2010). Expert systems are interactive and respond to questions, ask for clarifications, make recommendations, aid in decision making and give explanations.

RCM Definition

Reliability Centred Maintenance (RCM) is designed to minimise maintenance costs by balancing the higher cost of corrective maintenance against the cost of preventive maintenance, taking into account the loss of potential life (Crockera & Kumar, 2000)

Criscimagma (2013) views RCM as a logical framework for determining the optimum mix of applicable and effective maintenance activities needed to sustain the operational reliability of systems and equipment while ensuring their safe, economical operation and support

The "…logical and structured framework…" seem to suggest that RCM should be developed and executed in a consistent and systematic manner.

This matter shall be discussed later in this chapter where an attempt is made to explain how Expert systems can be applied to develop and execute the RCM programme systematically.

NASA (2008) Notes that the role of RCM has expanded from development of FMEA to:

- Sustainability
- Energy efficiency
- Commissioning, re-commissioning and retro-commissioning
- Maintainability
- Age Exploration
- Reliability analysis

Kirby (2012) in his white paper concurs with Criscimagna in that he asks about "How much maintenance is enough?" This question is answered by the optimisation of maintenance processes. There is a tendency by Maintenance Managers to do more maintenance than necessary in the Hydro-power generation industry to try and prevent equipment failure. This has not helped much given the phenomenon of maintenance induced failures and infant failures together with possible poor workmanship. The cost of outages as well as the cost of maintenance itself has resulted in the need to apply RCM to the hydro power generation industry as well. Whilst the definitions vary widely in the wording, they generally concur around RCM being logical, optimum and sustainable.

The RCM Processes

Moubray (1997) stresses that during the RCM process there are seven basic questions which should be asked about the asset or system under review. Mark G and Plucknette J (2008) stated the seven questions as follows:

1. What are the functions of the equipment or process?
2. How can it fail to provide the function?
3. What causes each functional failure?
4. What are the effects of each functional failure?
5. How does each failure impact the goals?
6. What action should be taken to predict and prevent each failure?
7. What action should be taken if a proactive task cannot be determined?

These questions can be answered by knowledge and experience acquired from archives of previous diagnosis processes of the equipment or system failure. Diagnostic expert systems are able to analyse various complex abnormal phenomena on equipment and can perform more detailed diagnosis automatically (Wang, 2003).

Decision Logic Tree

Figure 7 indicates the decision tree of the RCM process. This helps to make the appropriate maintenance strategy decision depending on the asset and various other factors.

RCM Programme Components

RCM is generally divided into four components:

1. Reactive Maintenance
2. Preventive Maintenance
3. Condition Based Maintenance
4. Proactive Maintenance

Figure 7. Decision logic tree
Kirby 2012.

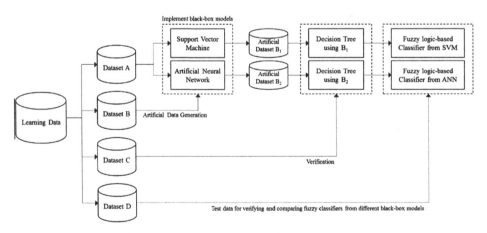

Figure 8. Representing relationships in a centrifugal pump system
Wang, 2003.

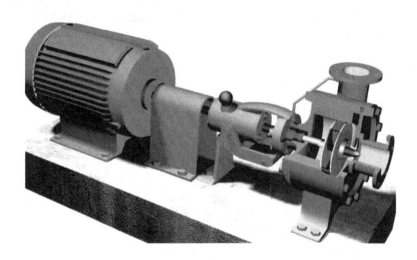

Reactive Maintenance

Swanson (2001) describes reactive maintenance as a firefighting approach to maintenance where equipment is allowed to run until failure. The failed component or equipment is repaired or replaced. Reactive maintenance is normally termed as quick fix in industry. This strategy has its pros and cons and is suitably applied on equipment of low priority equipment which is cheap and easy to replace but do not threaten the functionality of the system.

Faulty Tree Analysis

Rapid reaction and diagnosis to a faulty is a critical factor in returning plant to service. Expert systems have been developed to quickly diagnose failed equipment. Some of the systems use decision trees and fault tree analysis methods to analyse equipment failure, for instance a centrifugal pump system as illustrated in Figure 8.

Reliability Metrics

The reliability metric is a numerical figure that describes the reliability level associated with an item or component (the level at which this is classified will change depending upon the analysis). Currently MTBF is used as the basis of reliability predictions for items within a system (Relf, 1999). MTBF is the key reliability metric used for reliability. It is based on failure rates where MTBF is the reciprocal of failure rates. MTBF determines the preventive maintenance intervals.

Sources of MBTF

Historical information from CMMS

- Operator logs
- Parts usage
- Contractor records

$$Failure\ rate = \frac{Number\ of\ failures}{Total\ operating\ hours} = \frac{1}{MTBF} \qquad (7)$$

Inherent availability,

$$A = \frac{MTBF}{MTBF + MTTR} \qquad (8)$$

MTTR is the mean time to repair.

Inherent availability is the ideal availability where equipment is repaired as soon as it fails. However in real situations, there is need to trouble shoot and prepare before the actual repair or work is done on the system. The total time taken to do the actual work is known as Mean down time (MDT). It follows therefore that the operational availability Ao is given by:

$$A_0 = \frac{MTBF}{MTBF + MTTR} \qquad (9)$$

Reliability

$$Reliability = 1 - F(t) \qquad (10)$$

where F(t) is the probability that the system fails at time t

Figure 9. System reliability
NASA (2008).

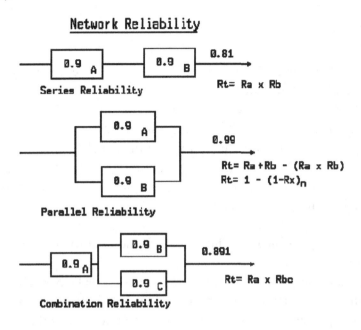

$$R(t) = e^{-t/_{MTBF}} = e^{-F(t)} \tag{11}$$

where R(t) is the reliability at time t

1. **System Reliability:** Systems often consist of equipment configured in networks which are serial, parallel or hybrid. Each equipment or subsystem presents its reliability and affects the overall reliability of the system.
2. **Series Configuration:** Overall

$$R = R_A * R_B * R_C \tag{12}$$

A series configuration has the disadvantage that the reliability is lower than the individual systems since the fractions are multiplicative.

3. **Parallel Configuration:** This is the configuration normally used in redundancy systems. One system is a duty and the other standby. This system is more reliable than a series set up.

The overall reliability is given by

$$R = 1 - (1 - R)^n \tag{13}$$

4. **Hybrid Configuration:** This set up combine the advantages of the series and parallel system.
5. **MTBF Critics:** Mobley (2002) thinks that the use of MTBF to plan maintenance usually results in either unnecessary repairs or catastrophic failure. This is probably due to the fact that MTBF is mainly based on statistical probability yet the equipment failure is randomly distributed and the MTBF is not constant. Dinesh et al (1999) in their paper listed the main drawbacks of MTBF as:

It is almost impossible to predict MTBF, if the time-to failure distribution is not exponential;

The methodology most widely used to predict MTBF and failure rate is based on the exponential distribution. This distribution is used to model failure times primarily because of its mathematical friendliness and the belief that Drenick's limit theorem is universally applicable rather than any scientific reason.

Maintenance Free Operating Period (MFOP)

There has been new development in the specification of reliability metrics emanating from the British Royal Air Force. The developments are meant to replace MTBF with MFOP as reliability metrics. The concept of the Maintenance Free Operating Period, or MFOP, was first proposed in 1996 by the UK Ministry of Defence as a means of aiding manufacturers of military aircraft to enable better operational planning capability, improved operational availability of aircraft and reduced running costs (Chew et al, 2007). MFOP based maintenance projects motivated by the need to guarantee airworthiness in the aviation industry have been implemented. Such projects include future offensive aircraft systems (FOAS), joint strike fighter (JSF) and ultra-reliable aircraft (URA) (Kumar, 2000).

MFOP is a period of operations during which the equipment should be able to carry out all its assigned missions without any maintenance action and without the operator being restricted in any way due to system faults or limitations (Relf et al, 1999). This new technique can be viewed as a warranty period throughout the equipment Life Cycle. Relf et al (1999) emphasises the key terms to focus on in this definition as 'the elimination of un-planned maintenance' and 'without any maintenance actions. There are other operator activities which are allowed during the maintenance free operating period like refuelling, tyre change and oil top up.

According to, Todinov (2003) the MFOP is essentially a period free from critical failures which require intervention for unscheduled maintenance. This means that the system is also allowed to carry defects as long the equipment functionality is retained. After each MFOP period, another period called the maintenance recovery period (MRP). During each MRP, the equipment is restored to a condition such that it is capable of completing the next MFOP, that is, the future assigned missions.

Trend Analysis

Failure data set should be tested whether the failure is a Homogenous Poisson Process (HPP) or Non-homogenous Poisson Process (NHPP). Shaalane (2012) describes a HPP as a process where the frequency of the number of failures in an interval of fixed length does not vary no matter when the interval is sampled and an NHPP as a process where the frequency of the number of failures in an interval of fixed length varies, at either an increasing or decreasing rate. The data is tested for a trend in the chronologically ordered data set, to check for increases or decreases in the inter arrival times of the data set.

Vlok (2011) and O'Connor (1995) state that the Laplace test can be first applied to a data set to compare the centroid of the observed arrival values with the midpoint of the period of observation.

The Laplace test tests the following hypothesis:

H_0: HPP
H_1: NHPP

Under H_0, it is assumed that the first n - 1 arrival times, $T_1; T_2;::::; T_{n-1}$, are uniformly distributed on (0; T_n). (Bartholomew, 1955).

The Laplace test statistic is as defined in Equation below:

$$U_L = \frac{\sum\limits_{i=1}^{n-1} T_{i-1} - T_{n/2}}{T_n \sqrt{\dfrac{1}{12(n-1)}}}$$

(14)

where

n = Number of failures,
Ti = i_{th} failure arrival time.

Shaalane (2012) indicates that $U_L \geq 2$ when there is strong evidence for reliability degradation, while if $U_L \geq -2$ this indicates a reliability improvement. Between $-1 \leq U_L \leq 1$, there is no evidence of a trend and it is therefore referred to as a noncommittal data set. In the last two cases where $2 > U_L > 1$ or $-1 > U_L > -2$, the Laplace test cannot provide indication with certainty that a trend is present in the data set or not. Lewis-Robinson tests can be used when the Laplace tests cannot be used with certainty (Wang and Coit, 2004)

The Lewis-Robinson test hypothesis is stated as:

H$_0$: renewal process
H$_a$: not a renewal process

U_{LR} the test statistic is given by Equation below:

$$U_{LR} = \frac{U_L}{CV}$$

(15)

where

U_L = The Laplace test statistic
CV = The coefficient of variation of inter arrival times

Non-Repairable System Analysis (NRS)

A Non-Repairable system is that which is discarded once it fails. (Shaalane, 2012)

If after the Laplace trend test has been applied to the distribution data, and the inter-arrival times of the data yield no trend, the data set should be analysed using non- repairable systems theory. O'Connor et al. (1995),

Vlok (2011) and Montgomery et al. (2010) recommended the use of the Weibull distribution in reliability modelling. Equation 2.10 is the Probability Density Function (pdf) which provides the probability of system failure at instant, x.

$$f(x) = \frac{\beta}{\eta} \left(\frac{x}{\eta}\right)^{\beta-1} . \exp\left(-\left(\frac{x}{\eta}\right)^{\beta}\right) \tag{16}$$

where

β = The shape parameter
η = The scale parameter
$\beta > 0$ and $\eta > 0$

β and η are used to calculate the MFOP of the system with the given data. Maximising the likelihood in equation 2.11 would give the Weibull parameters β and η:

$$InL\left(X_i, \theta\right) = \sum_{i=1}^{m}\left[In\frac{\beta}{\eta} + (\beta-1)In\frac{X_i}{\eta}\right] - \sum_{j=1}^{r}\left(\frac{X_j}{\eta}\right)^{\beta} \tag{17}$$

The cumulative probability function can be derived from equation above; this is shown in equation below.

$$F(x) = 1 - \exp\left(-\left(\frac{x}{\eta}\right)^{\beta}\right) \tag{18}$$

Equation 19 shows the reliability function for the Weibull distribution

$$R(x) = \exp\left(-\left(x/\eta\right)^{\beta}\right) \tag{19}$$

1. **Calculation of MTBF:** For a non-repairable system, the *historic* MTBF can be interpreted as the mean of the observed failure data as shown by Equation below:

$$MTBF = \frac{\sum X_i}{m} \tag{20}$$

where X_i is the inter-arrival times of failures and m is the total number of observed failures.

The *future or predicted* MTBF can also be found, as shown in equation below (Vlok, 2011)

$$E\left[X_{r+1}\right] = \frac{\int_0^\infty x \cdot f(x)dt}{\int_0^\infty f(x)dt} \tag{21}$$

2. **Calculation of MFOP:** It is practically not possible to guaranteed a 100% MFOP and hence the need for Maintenance Free Operating Period Survivability (MFOPS).

Kumar et al. (1999), Long. (2009) and Chew (2010) defines MFOPS as the probability that the part, subsystem, or system will survive for the duration of the MFOP, given that it was in a state of functioning at the start of the period. MFOPS is given by Equation below.

$$MFOPS(t_{mf}) = \exp\left(\frac{t^\beta - (t - t_{mf})^\beta}{\eta^\beta}\right) \tag{22}$$

where

η = The scale parameter
β = The shape parameter of the Weibull distribution.

Rearranging Equation above, the MFOP length for a given confidence (MFOPS), Equation below is obtained:

$$(t_{mf}) = \left[t^{\beta} - \eta^{\beta} In \left(MFOPS \left(t_{mf} \right) \right) \right]^{1/\beta} - t \tag{23}$$

The maximum achievable MFOP at a given MFOPS is given by, Equation below

$$MFOP = \eta * \left(In(\frac{1}{MFOPS'}) \right)^{1/\beta} \tag{24}$$

This equation is used to calculate MFOP which symbolises the life of a non-repairable system.

Repairable Systems Analysis (RS)

Shaalane (2012) defines an RS as a system that can be reinstated again to perform all of its functions, other than a complete replacement of the system.

NHPP describes a process where the rate at which events occur is not constant and the power law NHPP is used to model the repairable system, given in Equation 25 as:

$$\rho_1 = \lambda \delta t^{\delta - 1} \tag{25}$$

The parameters, δ and λ are constants for a specific system.

Expected number of failures, N, between any two points in time, t_1 and t_2 are given by Equation 26.

$$E[N(t_1 - t_2)] = \lambda(t_2^{\delta} - t_1^{\delta}) \tag{26}$$

Vlok (2011) recommends that the parameters for the power law NHPP be estimated using the least-squares method, this being the difference between the observed number of failures and the number of failures expected by $\rho_1(t)$, which should be minimised.

Two types of events are possible in the analysis, a failure or a planned outage (suspension). For a failure event, Equation 27 is applicable.

$$\min(\tilde{\lambda}, \tilde{\delta}) : \sum_{i=1}^{r} \left[E\left[N\left(0 \to T_i\right)\right]\right] - N(0 \to T_i)^2 \tag{27}$$

For a planned (suspended) event type, Equation 28 is applicable.

$$\min(\tilde{\lambda}, \tilde{\delta}) : \sum_{i=1}^{r-1} \left[E[N(0 \longrightarrow T_i)] - N(0 \longrightarrow T_i)^2 \right] \tag{28}$$

Now that the parameter, δ and λ, can be estimated, the reliability of the system can be determined from t_1 to t_2 by using the equation 29:

$$R(t_1 \longrightarrow t_2) = e^{-\lambda(t_2^\delta - t_1^\delta)} \tag{29}$$

1. **Calculation of MTBF:** For a repairable system, the *historic* MTBF can be calculated by using Equation below.

$$MTBF_{\rho 1}(t_1 \longrightarrow t_2) = \frac{t_2 - t_1}{\lambda(t_2^\delta - t_1^\delta)} \tag{30}$$

The future MTBF provides a better representation of the current system state. If the system is put back in operation after the last recorded failure, then the next failure of the system can be predicted, $(r+1)^{th}$. Equation 31 can be used to calculate the residual life of the system from the last recorded event.

$$E[(T_{r+1} t = T_r) = (\frac{1 + \lambda T_r^\delta}{\lambda})^{1/\delta} \tag{31}$$

2. **Calculation of MFOP:** The probability of surviving (MFOPS) t_{mf} units of time is given in Equation below:

$$MFOPS(t_{mf}) = e^{-\lambda((t_{mf} + T_r)^\delta - (T_r)^\delta} \tag{32}$$

Figure 10. MFOP and MTBF calculation algorithm

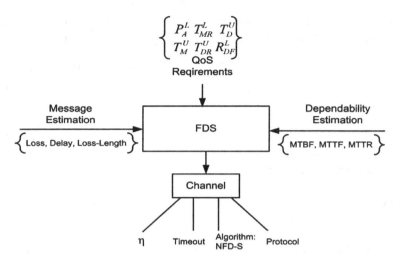

where

T_r = The global time unit of the last known failure event and the parameters λ and δ = Constants found by the least squares method.

MFOP and MTBF Calculations

The flow chart in Figure 10 details the algorithm used to process the data set. The equations detailed above can be formulated in Ms Excel to determine the parameters of both reparable and non-repairable systems. The failure data set is tested for trend using Laplace tests and Lewis-Robson tests through to the determination of MFOP, MTBF and the reliability of the system. Ms Excel also serves as a database for the failure data collected from the field.

REFERENCES

Blanchard & Fabrycky. (1990). *Systems engineering and analysis.* Englewood Cliffs, NJ: Prentice Hall.

Heng, A., Zhang, S., Tan, A. C., & Mathew, J. (2008). Rotating machinery prognostics: State of the art, challenges and opportunities. *Mechanical Systems and Signal Processing*, 23(3), 724–739.

Jardine, A. K., Lin, D., & Banjevic, D. (2006, October). A review on machinery diagnostics and prognostics implementing condition-based maintenance. *Mechanical Systems and Signal Processing*, 20(7), 206–217.

Lee, J., Ni, J., Djurdjanovic, D., Qiu, H., & Liao, H. (2006, February 9). Intelligent prognostics tools and e-maintenance. *Computers in Industry, 57*(6), 476-489.

Liu, J. (2008). An extended wavelet spectrum for bearing fault diagnostics. *IEEE Transactions on Instrumentation and Measurement, 57*(12), 2801–2812. doi:10.1109/TIM.2008.927211

Moubray. (1997). *Reliability-Centered Maintenance*. New York: Industrial Press.

Okah-Avae. (1981). *Condition monitoring – A maintenance management strategy for industrial machinery*. Benin: University of Benin-City.

Shayeri. (2007). Development of Computer-Aided Maintenance Resources Planning (CAMRP): A Case of Multiple CNC Machining Centers. *Robotics and Computer Integrated Manufacturing*, 614-623.

Sondalini. (2014, December 20). *Lifetime Reliability Solutions*. Retrieved March 3, 2015, from Now get world class asset management, maintenance and reliability in your company the Plant and Equipment Wellness Way: http://www.lifetime-reliability.com/

Tao, Chan, & Wang. (2013). An approach to performance assessment and fault diagnosis for rotating machinery equipment. *EURASIP Journal on Advances in Signal Processing,* 738-746.

Ugechi, Ogbonnaya, Lilly, Ogaji, & Probert. (2009). Condition-Based Diagnostic Approach for Predicting the Maintenance Requirements of Machinery. *Scientific Research Journal*, 177-187.

Zio. (2007). Soft Computing Methods applied to condition monitoring and fault diagnosis for maintenance. In *Summer Safety and Reliability Seminars* (pp. 43-52). Gdansk/SopotJelitkowo.

KEY TERMS AND DEFINITIONS

ESMDT: Engineering, science, and management defence training.

RCM: Reliability-centred maintenance.

Chapter 3
Expert Systems and Fuzzy Logic

ABSTRACT

An expert system is a system that employs human experience or knowledge captured in a computer to solve problems that ordinarily require human expertise. They may or may not have a learning component. Expert systems are a branch of Artificial intelligence. Truemper describes an expert system as an intelligent system which in an interactive setting asks a person for information and, based upon the response, draws conclusions or gives advice. Problems tend to be solved using heuristics (rules of thumb) or approximate methods or probabilistic methods which, unlike algorithmic solutions, are not guaranteed to result in a correct or optimal solution. The authors go further to clarify that expert systems usually have to provide explanations and justifications of their solutions or recommendations in order to convince the user that their reasoning is correct.

EXPERT SYSTEMS

Expert systems have evolved over the years and have been applied to medical diagnoses systems, robotics, speech recognition and language processing among others. It can be noted from Wang et al (2003) that expert systems have also been applied in maintenance as condition monitoring and diagnostic systems. Expert systems are different from automation or step-by step

DOI: 10.4018/978-1-5225-3244-6.ch003

programming or conventional computing. In the case of Expert systems, Wang et al (2003) points out that the computer is given knowledge about the subject area plus some inference capability. Expert systems are not just a set of algorithms or some mathematical formulae software based on symbolic representation and manipulation. A symbol, letter, sentence or number is used to represent objects like things, ideas, events or statement of facts. Giarratano et al stated the analogy of the two systems as:

- Conventional Computer Algorithms

$$Algorithms + Datastructures = Programs \tag{1}$$

- Expert Systems

$$Knowledge + Inference = ExpertSystems \tag{2}$$

Expert System Architecture

Figure 1 below shows the system architecture. The data is input by the user or some sensors through the user interface. The user interface has an interviewer component and an explanation facility. The interviewer component controls the dialog with the user and allows measured data to be read into the system. The system for instance asks the user a series of questions, or read a file containing a series of test results (Bullinaria, 2005).

The output of the system is advice, suggestions and explanations. Wang et al (2003) realises that expert systems should give guidelines or instructions on remedy measures on failed equipment, just like the way a human Engineer would do.

The explanation facility provides the ability to trace the inference paths. Bullinaria et al (2005) explains that the explanation facility gives the system's solution, and provides the user with information about its reasoning process. It might output the conclusion, and also the sequence of rules that was used to come to that conclusion. It might instead explain why it could not reach a conclusion.

Figure 1. Expert system architecture
Aziz, 2005.

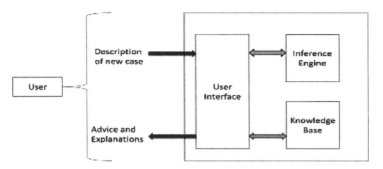

Knowledge Base

This component of Expert system is necessary for understanding, formulating, and solving problems (Wang, 2003). Expert system maintains the expert's domain knowledge in a module known as knowledge base. According to Aziz (2005) the Expert system uses rules technique to code the knowledge in the knowledge base. A rule is an IF/THEN structure that logically relates information contained in the IF part to other information contained in the THEN part.

Inference Engine

This is the brains behind the system. It is the control structure and interprets the rules. Wang et al (2003) describes the inference engine as the component which give direction of how to use the system knowledge. It can be generally described as the search method in an expert system. Wikipedia describes an inference engine as applying logical rules to the knowledge base and deduced new knowledge. This process would iterate as each new fact in the knowledge base could trigger additional rules in the inference engine ('Inference Engine' n.d.).

Two Inference mechanisms are generally used. These mechanisms are forward chaining which is a data driven approach and backward chaining which is a goal driven approach (Wang et al, 2003)

THE FUZZY EXPERT SYSTEM PROCESS

Fuzzy systems theory is best suited in handling uncertainties, ambiguities, and contradictions, typical of those found in an ERCM system. Figure 2 shows the fuzzy processes.

- **Fuzzification** is a process of converting inputs into information that the inference mechanism can easily use to activate and apply rules.
- **Defuzzification** is the process of converting the conclusions of the inference mechanism into outputs.

A Production System Cycle

A typical Expert system flow diagram can be represented as shown in Figure 3.

EXPERT SYSTEM SHELLS

Shells are tools for building expert systems that provide knowledge representation facilities and inference mechanisms. Knowledge Pro is an

Figure 2. The Fuzzy Expert System Process

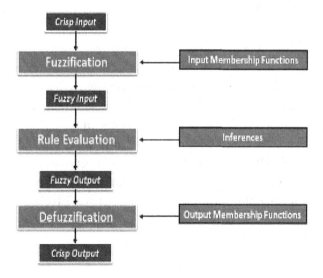

51

Figure 3. A production system cycle
Kasabov (1996).

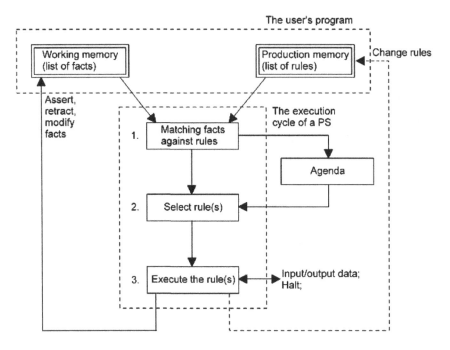

expert system shell which is useful in developing an expert system. Knowledge Pro runs on windows platform. It is also important to note that Knowledge Pro application can exchange data with Ms Excel. Expert systems can be programmed using 4th generation languages like Lotus 1-2-3, C++ and COBOL. (Wang, 2003).

Whilst this author recognise the works done on expert systems as applied to maintenance, it is clear that more research is required around Expert Reliability Centred Maintenance Systems, as applied to hydro-power stations. Expert systems are ideal where human expertise is not always available and the problem requires symbolic reasoning. This makes Reliability Centred Maintenance a suitable candidate in this era of brain drain and skill shortage in Zimbabwe. Most RCM reasoning has some degree of uncertainty while other concepts are crisp. This phenomenon makes both fuzzy and non-fuzzy expert systems applicable in the development of the ERCM.

FUZZY LOGIC

Most companies are using what is termed smart technologies in reducing machinery breakdown (Innovolt, 2014). The term control is generally defined as a mechanism used to guide or regulates the operation of a machine, apparatus or constellations of machines and apparatus. Often the notion of control is inextricably linked with feedback: a process of returning to the input of a device a fraction of the output signal. Feedback can be negative, whereby feedback opposes and therefore reduces the input, and feedback can be positive whereby feedback reinforces the input signal (Passino K and Yurkovich S, 1998).

Figure 4 shows a typical feedback conventional control mechanism and it controls the error if it arises. In the figure,

M = Machinery,
F = Feedback,
s = Signal of the error,
i = Input,
u = Control signal or non-linear function,
y = Output

The control signal (u) can either be; proportional to the error, proportional to both the magnitude of the error and the duration of the error or lastly can also be proportional to the relative changes in the error values over time. There are also constants in this setup of feedback which are the proportional constant (K_p), derivative constant (K_D) and the integral constant (K_I).

According to (Babuska and Mamdani, 2008); these constants can be linked as follows;

Figure 4. Feedback conventional control mechanism

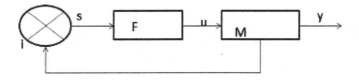

$$u(t) = K_p s(t) + K_I \int_0^t s(\tau)d\tau + K_D \frac{ds(t)}{dt} \tag{3}$$

With equation 2 deductions of the following is carried out and concluded.

In the case of classical operations of process control one has to solve the non-linear function u. Furthermore, it is very important that one also finds the proportionality constants (PID). In the case of fuzzy controller, the non-linear function is represented by a fuzzy mapping, typically acquired from human beings (Babuska and Mamdani, 2008). The conventional controller used to work as for the general PID but it will face some challenges in case of robotics section. This is where rules and laws are generated, the IF THEN ELSE rules and put in the fuzzy logic software. Figure 5 shows the arrangement of components in fuzzy logic controller.

The fuzzy controller uses intelligent sensors that react faster if any error or fault occurs. Figure 6 shows the fuzzy controller with a sensor and the

Table 1. What the PID means

Value (PID)	Determines Reaction to the
Proportional (K_p)	Current error
Integral (K_I)	Sum of recent errors
Derivative (K_D)	Rate at which the error has been changing

Babuska and Mamdani, 2008.

Figure 5. A fuzzy logic based controller (FLC)
Babuska and Mamdani, 2008.

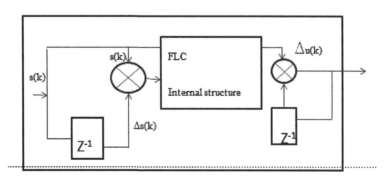

Figure 6. Fuzzy controller
Babuska and Mamdani, 2008.

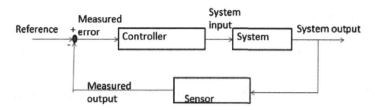

controller. A fuzzy controller by definition is a device that intends to model some vaguely known or vaguely described processes (Babuska and Mamdani, 2008).

Fuzzy logic has basically two types of controllers which are the Mamdani and Takagi-Sugeno (Yager R R and Filev D P, 1994). In this manner the researcher will focus on the Takagi-Sugeno-Kang method which uses (Supervisory Control and Data Acquisition) SCADA for online monitoring.

The controller can be used with the process in two modes: Feedback mode when the fuzzy controller will act as a control device; and feed forward mode where the controller can be used as a prediction device (Yager R R and Filev D P, 1994). A controller is implemented using an algorithm. This controller is to be used in this research for maintenance duties in CBM.

Advantages of Sugeno and Mamdani Method

Advantages of the Sugeno Method

- It is computationally efficient.
- It works well with linear techniques (e.g., PID control).
- It works well with optimization and adaptive techniques.
- It has guaranteed continuity of the output surface.
- It is well suited to mathematical analysis (Babuska and Mamdani, 2008).

Table 2. Controllers under fuzzy logic

Controller Type	Typical Operation
Mamdani (linguistic) controller with either fuzzy or singleton consequents.	Direct close-loop controller
Takagi-Sugeno (TS) or Takagi-Sugeno-Kang controller	Supervisory controller – as a self-tuning device

Babuska and Mamdani, 2008.

Advantages of the Mamdani Method

- It is intuitive.
- It has widespread acceptance.
- It is well suited to human input (Babuska and Mamdani, 2008).

Fuzzy inference system is the most important modeling tool based on fuzzy set theory. The FISs are built by domain experts and are used in automatic control, decision analysis, and various other expert systems.

- Control engineers have traditionally relied on mathematical models for their designs. If the system is complex the mathematical modeling is less effective. In general the benefits of adopting fuzzy control are:
  ◦ Fuzzy controllers are more robust than PID controllers because they can cover a much wider range of operating conditions than PID can, and can operate with noise and disturbances of different natures, that is why the researcher is using them in this work.
  ◦ Developing a fuzzy controller is cheaper than developing a model-based or other controller to do the same thing.
  ◦ Fuzzy controllers are customizable, since it is easier to understand and modify their rules, which not only use a human operator's strategy but also are expressed in natural linguistic terms.

It is easy to learn how fuzzy controllers operate, and how to design and apply them to a concrete application.

Application of Fuzzy Logic in Washing Machines

Washing machines also known a laundry machines are machines used to wash clothes and sheets. They reduce the effort put in by the customer to brush and wash the dirty clothes. Different clothes need different times for washing this depending on the amount of dirt on the clothes and the quality of the material. Ordinary washing machines without fuzzy logic control do not determine automatically the precise amount of time required for each item. Therefore, the use of fuzzy logic control presents the idea of controlling the times used for washing. The process for controlling the time required for washing uses the principle of fuzzy logic whereby non-precise inputs from the sensors are subjected to the fuzzy arithmetic. A crisp value is then obtained of the washing time. Sensors are used to automate the process and they detect parameters for example the type and degree of the dirt and the volume of the clothes. This data is then used to determine the amount of time for washing. A fuzzy logic controlled washing machine gives the exact time required for washing.

Two inputs are taken by the FLC and these inputs are the type of dirt and the degree of dirt. These inputs are then processed and an output of the washing time is then given as shown in Figure 7. The sensor determines the degree of dirt by analysing the transparency of the water used for washing. The sensors also determine the type of dirt by analysing the amount of time taken to reach saturation. The saturation point is the point whereby there is no further change in the colour of the wash water (Alhaddad, 2013). The FLC makes decisions based on the rules stored in the database. The output is derived from the following set of rules:

- If dirtiness of clothes is Large and type of dirt is Greasy then wash time is Very Long;
- If dirtiness of clothes is Medium and type of dirt is Greasy then wash time is Long;
- If dirtiness of clothes is Small and type of dirt is Greasy then wash time is Long.

Figure 7. Process block diagram

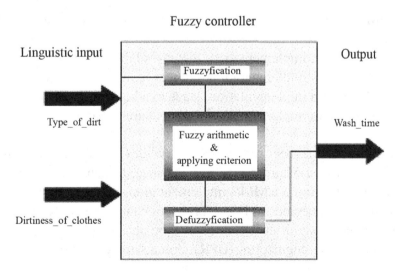

57

A Synergy of Fuzzy Logic and Condition Monitoring Done Up to This Period.

Unlike the usual CBM as seen above, in this area of synergizing fuzzy and CBM is to actually identify as and when a fault occurs. In the general CBM, a fault is identified when the apparatus are put in place and identify that a fault has already occurred. This fuzzy logic is there to show exactly when it happen and protect other machinery and reduce time wastes in the company. Fuzzy logic is among the knowledge-based techniques to address the fault detection problem (Iserman, 1998). Several researchers have proposed fault detection and diagnosis approaches based on a fuzzy system (Schneider, 1993). Neural networks have been proposed to solve the problem but they are only suitable for building fuzzy systems with a relatively small number of numerical variables. The trapping of the local optimal in the learning process is the main weakness of using neural networks (Yuan Y F and Zhuang H J, 1996). (Mostafa M, Payam M, Hatami A M, Alireza G and Allah K H, 2013), highlighted that maintenance happened in periods i.e., first period of evaluation is time before second world war, second evaluation period up to the 1970s and the third evaluation period which is currently operating as of now. In their research they realized that most companies are doing Computerized Maintenance Management Systems (CMMS) which is more of CBM but also has some problems as we are moving to highly advanced machinery which needs intelligence as in this research. (Dunn, 1997) has implied the advantages of CMMS as:

- CMMS system implementation, increases the equipment availability time
- CMMS system implementation, increases labor efficiency
- CMMS system implementation, reduces maintenance costs
- CMMS system implementation, leads to a better stock control
- CMMS system implementation reduces bureaucracy.

More recently, it has been recognised that, while CMMS can be a significant enabler of improved Maintenance performance, to achieve the maximum possible benefits from a CMMS implementation, business processes must be formally re-engineered, and work practices changed in a coordinated and planned manner if significant change is to be achieved. Thus was born the Business Process Re-engineering (BPR) approach to Systems Implementation, which focuses on both the Technology and the Business Processes involved in a CMMS implementation (Dunn, 1997).

However, even this BPR approach has, in recent years, become the subject of some debate, with many suggesting that this approach is not wholly successful in achieving long-term, sustainable benefits. While most of our attention is focused on the technology associated with the CMMS software and hardware, we should not ignore the opportunities that may exist by utilising other technologies around our plant and equipment, and linking these to our CMMS. These days, the greatest benefits that are being obtained from CMMS implementations being gained by linking Engine Management systems and Process Control Systems to our Maintenance systems. These systems can monitor equipment performance, and give us an early warning that a piece of equipment may require some maintenance attention (Dunn, 1997). With this in mind the author thought of coming up with a fuzzy logic approach to machine monitoring.

(Berger, 2010) conducted a study about implementation of CMMS, and suggested that the barriers to implant CMMS are:

- Lack of management support for the implementation of CMMS systems
- Employee resistance in the CMMS system
- Poor planning for the implementation of automated systems
- Lack of appropriate training software and its capabilities to employees.

According to Berger that shows poor training of advanced tools to assist in maintenance so the use of fuzzy logic if highly understood can be a very good solution. (Wu By C S, Polte T and Rehfeldt D, 2001) introduced a fuzzy logic system that is able to recognize common disturbances during automatic gas metal arc welding (GMAW) using measured welding voltage and current signals. A statistical method was employed to process the captured transient raw data, and the probability density distributions (PDDs) and the class frequency distributions (CFDs) were obtained. Based on the processed data (PDD values of welding voltage and current and CFD values of the short-circuiting time), the system automatically generates fuzzy rules and membership functions of linguistic variables, conducts inference and defuzzification, and completes the evaluation process without further processes. (Ahmadi H, Moosavian A and Khazaee M, 2012) did a fault diagnosis on rotating machinery using fast Fourier transforms (FFT) and Matlab 7.6.

(Momeni M, Fathi M R, Zarchi M K and Azizollahi, 2011) differentiated maintenance strategies for different equipment. The selection of different maintenance strategies is a typical multiple criteria decision making (MCDM) problem. Considering the imprecise judgments of decision makers, now the fuzzy TOPSIS is used for the evaluation of different maintenance strategies. Fuzzy TOPSIS favors preventive maintenance approach to machinery.

(Ahmadi H, Moosavian A and Khazaee M, 2012) Research proved that Support Vector Machines (SVM) can reduce misalignments in machinery when vibrating. It is an intelligent control of CBM. Vibration analysis is one of the main techniques used to the non-destructive diagnosis and identification

Figure 8. The schematic of the diagnostic method
By (Ahmadi H, Moosavian A and Khazaee M, 2012).

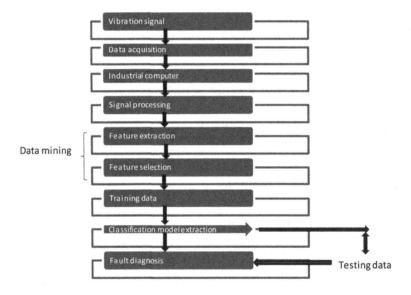

Figure 9. Data acquisition system (SVM) equipment
By (Ahmadi H, Moosavian A and Khazaee M, 2012).

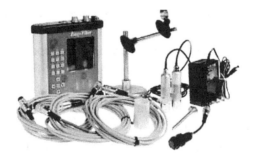

of various defects in rotary machines. Vibration analysis provides early information about progressing malfunctions for future monitoring purpose.

(Jolandan S G, Mobli H, Ahmadi H, Omid M and Mohtasebi S S, 2012) came up with an experiment to use fuzzy logic to find faults in a gearbox of a tractor.

The parts in Figure 11 represent the data in Table 3.

A combined classification tree (J48 algorithm) and fuzzy inference system (FIS) have been presented to perform fault diagnosis of a gearbox. The implementation of J48-FIS based classifier requires two consecutive steps. Firstly, method Correlation based Feature Selection (CFS) and the J48 algorithm is utilized to select the relevant features in the data set obtained from feature extraction part. The output of the J48 algorithm is a decision tree

Figure 10. Flowchart of fault diagnosis system
By (Jolandan S G, Mobli H, Ahmadi H, Omid M and Mohtasebi S S, 2012).

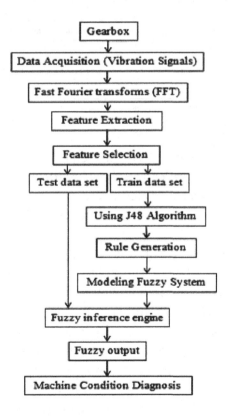

Figure 11. Experimental test bed as done
By *(Jolandan S G, Mobli H, Ahmadi H, Omid M and Mohtasebi S S, 2012).*

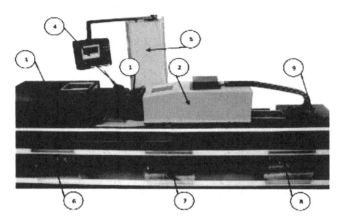

Table 3. Components that make up the tractor system

1.	Gearbox	6.	Electrical and pneumatic valves
2.	Electrical motor	7.	Hydraulic system
3.	Load establisher system	8.	Lubrication system
4.	Controller keys and touch panel	9.	Power cylinder to move electromotor
5.	Electrical and electronic Toolbox		

Jolandan S G, Mobli H, Ahmadi H, Omid M and Mohtasebi S S, 2012.

that is employed to produce the crisp if-then rule and membership function sets. Secondly, the structure of the FIS classifier is defined based on the obtained rules, which were fuzzified in order to avoid classification surface discontinuity (Jolandan S G, Mobli H, Ahmadi H, Omid M and Mohtasebi S S, 2012).

The whole idea in this research is to automate our machinery to avoid breakdowns. (Rzevski, 1995) highlighted the use of automation with intelligence and referred to it as high automation.

According to (Rzevski, 1995) this is more or less that is required to be done by the researcher in machines to avoid machine breakdowns but will only use the intelligent tool as fuzzy logic. Traditionally maintenance procedures in industry follow two approaches as follows. The first one is to perform fixed time interval maintenance, where the engineers take advantage of slower production cycles to fully inspect all aspects of the machinery. The second is to simply respond to the plant failure as and when it happens. However,

Figure 12. Trends of automation

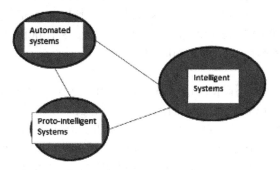

making use of today's technology, new scientific approach was becoming possible for maintenance management. Recently soft computing techniques such as expert system, neural network, fuzzy logic, adaptive neural fuzzy inference system, genetic algorithm etc. have been employed to assist the diagnostic task to correctly interpret the fault data. These techniques have gained popularity over other conventional techniques. The area of condition monitoring and faults diagnostic of electrical drives is essentially related to a number of subjects, such as electrical machines, methods of monitoring, reliability and maintenance, instrumentation, signal processing and intelligent systems (Dash, 2010). In the present work, several soft computing techniques have been presented for the detection and location of stator inter-turn short circuit fault in the stator winding of an induction motor.

- Discrete wavelet transform approach is successfully used to detect and locate stator interturn short circuit fault together with identification the severity of this fault in the stator winding of an induction motor. The same approach can be extended to identify the other faults such as bearing fault, rotor broken bar fault, and eccentricity related fault of an induction motor.

- Neural network ensembles are receiving increasing attention in the field of research such as control, fault diagnosis, decision making, identification, robotics etc. due to their learning and generalization abilities, nonlinear mapping, and parallelism of computation. However, for solving fault detection problems the neural networks may get stuck on a local minimum of the error surface, and the network convergence rate is generally slow. A suitable approach for overcoming these disadvantages is the use of wavelet functions in the network structure. Wavelet function is a waveform that has limited duration and an average value of zero. A wavelet neural network has a nonlinear regression structure that uses localized basis functions in the hidden layer to achieve the input-output mapping.

The integration of the localization properties of wavelets and the learning abilities of neural network results in the advantages of wavelet neural network over neural network for the detection and location of an inter-turn short circuit fault in the stator winding of an induction motor (Dash, 2010). This shows his research in terms of artificial intelligence and further work is still required in intelligent control of machinery. This led the researcher in this topic to come up with the fuzzy logic system.

Probability based analysis usually requires more information about the system than is known, such as mean failure rates or failure rate distribution which commonly results in dubious assumption about the original data. However, fuzzy logic provides necessary requirements in handling with imprecise and uncertain information in more consistent and logical manner (Kumar A, Sharma S P and Kumar D, 2007). Among the inexact reasoning methods known so far, fuzzy methodology (FM) acts as one of the most viable and effective tool.

Fuzzy has gained a lot of ground as it is used in fault diagnosis (Mustapha F, Sapun S M, Ismail N and Mokhtar A, 2004), structural reliability (Savoia, 2002) and (Wang W L, Pan D and Chen H M, 2006), human reliability (Konstandinidou M, Nivolianitou Z, Kiranoudis C and Markatos N, 2006), safety and risk engineering (Mostafa M, Payam M, Hatami A M, Alireza G and Allah K H, 2013) and (Guimaraes A C F and Lapa C M F, 2007) and quality control (Khan M K and Hafiz N, 1999) and (Sharma R K, Kumar D and Kumar P, 2007). This is what made the author to do the research using fuzzy logic. (Kumar A, Sharma S P and Kumar D, 2007) developed a system that may help maintenance engineers to analyze and predict the system behavior in robots. An attempt has also been made to deal with imprecise, uncertain dependent information related to system performance. Various reliability parameters (such as failure rate, repair time, mean time between failures, availability, reliability and expected number of failures) were computed to predict the system behavior in objective terms and it is concluded that in order to improve the availability and reliability aspects, it is necessary to enhance the maintainability requirement of the system. Fuzzy logic, fault tree analysis and Petri Net sets were used in the analysis.

(Tao, Chen, Chan and Wang, 2013) came up with a method called Fisher discriminant analysis (FDA) and Mahalanobis distance (MD) applied to rotating machinery fault diagnosis, which is further extended to performance assessment and fault detection.

SOME MAJOR COMPANIES PRACTICING AUTOMATION TO REDUCE MACHINE FAILURE

Automation refers to the use of computers and other automated machinery for the execution of business-related tasks. Automated machinery may range from simple sensing devices to robots and other sophisticated equipment. Automation of operations may encompass the automation of a single operation or the automation of an entire factory (Advameg, Inc., 2015). Although automation can play a major role in increasing productivity and reducing costs in service industries—as in the example of a retail store that installs bar code scanners in its checkout lanes—automation is most prevalent in manufacturing industries. In recent years, the manufacturing field has witnessed the development of major automation alternatives. Some of these types of automation include:

- Information Technology (IT)
- Computer-Aided Manufacturing (CAM)
- Numerically Controlled (NC) equipment
- Robots
- Flexible Manufacturing Systems (FMS)
- Computer Integrated Manufacturing (CIM)

In this context the researcher is going to make use of FMS, Robots and CIM where fuzzy will be the pivot of study. Industry accounts for more than 40 percent of worldwide energy consumption. Of this, more than 65 percent of industrial power demand comes from electric motor-driven systems. Therefore, reducing the energy use of motor-driven systems represents a significant opportunity for industrial energy savings (Koditek F, 2012). Controlling motor speeds can contribute significantly to energy savings; however, the same intelligence through automation that allows motor speeds to be controlled successfully can be used in other industrial applications as well that is to be done by the author here.

HUMAN MACHINE INTERFACE (HMI)

Also known as an HMI. An HMI is a software application that presents information to an operator or user about the state of a process, and to accept and implement the operators control instructions. Typically information is displayed in a graphic format (Graphical User Interface or GUI). An HMI is often a part of a SCADA (Supervisory Control and Data Acquisition) system (Subnet Solutions Inc., 2015). The Human Machine Interface (HMI) includes the electronics required to signal and control the state of industrial automation equipment. These interface products can range from a basic LED status indicator to a 20-inch TFT panel with touchscreen interface. HMI applications require mechanical robustness and resistance to water, dust, moisture, a wide range of temperatures, and, in some environments, secure communication (Atmel, 2015).

(Human Machine Interface) The user interface in a manufacturing or process control system. It provides a graphics-based visualization of an industrial control and monitoring system. Previously called an "MMI" (man machine interface), an HMI typically resides in an office-based Windows

Figure 13. HMI devices

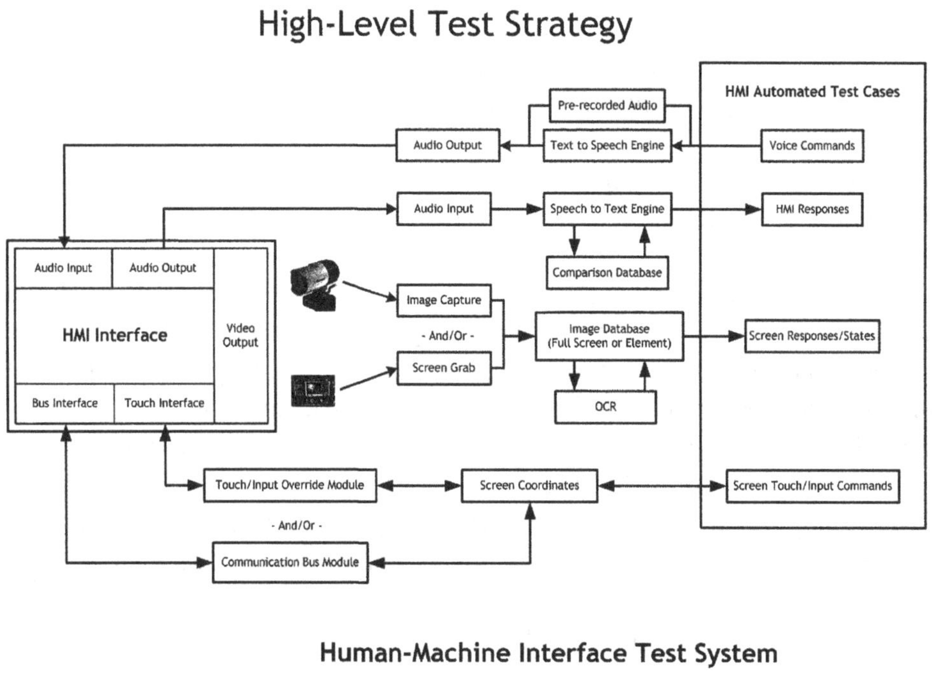

High-Level Test Strategy

Human-Machine Interface Test System

computer that communicates with a specialized computer in the plant such as a programmable automation controller (PAC), programmable logic controller (PLC) or distributed control system (DCS) (Encyclopaedia, 2015).

Input devices such as mice and joysticks to facilitate user input to a computer executing program code are well known in the art. A user manipulating an input device of this nature is able to interact with the software application being executed by the computer. Although these input devices are common, touch sensitive panels have also been considered to enable users to interact with software applications being executed by computers (USA Patent No. 7,859,519 B2, 2010). There are different types of HMIs used globally. A machine is an apparatus using or applying mechanical power, having several parts each with a definite function and together performing certain kinds of work (Charwat H. J, 1992).

Programming of these interfaces is done through system compatible software. Two types of ways are provided by the user interface by which the user can interact with the system. Which are; Input, which allows users to manage a system and Output, which allows the system to show the effects of the users' operation. The main types of human machine interfaces are:

- Graphical user interfaces (GUI)
- Voice user interfaces (VUI)
- Command line interfaces (CLI)
- Touch Interfaces (TI)

These can be used to monitor the plant and prevent machine breakdowns so often as what is required by the researcher.

Graphical User Interfaces (GUI)

Software that works at the point of contact (interface) between a computer and its user, and which employs graphic elements (dialog boxes, icons, menus, scroll bars) instead of text characters to let the user give commands to the computer or to manipulate what is on the screen (WebFinance, Inc., 2015).

Voice User Interfaces (VUI)

Today's enterprise organizations have two competing goals: they must provide their customers with a consistently high-level of customer service while simultaneously achieving cost savings. Speech-enabled over-the-phone

Figure 14. Example of automation and control system

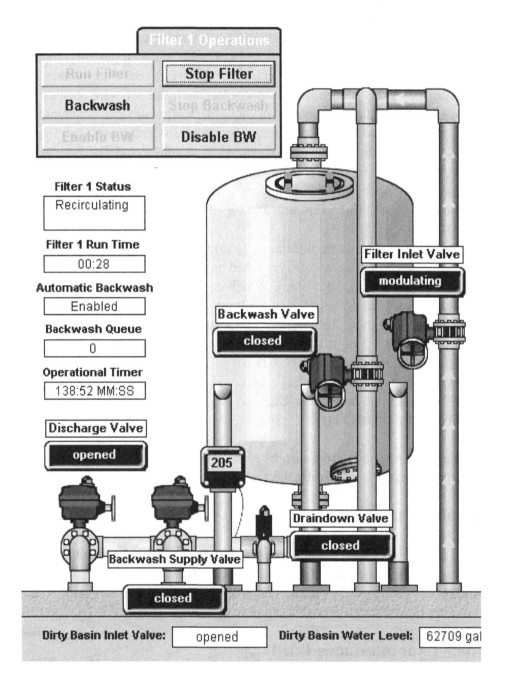

Figure 15. GUI in CAD system

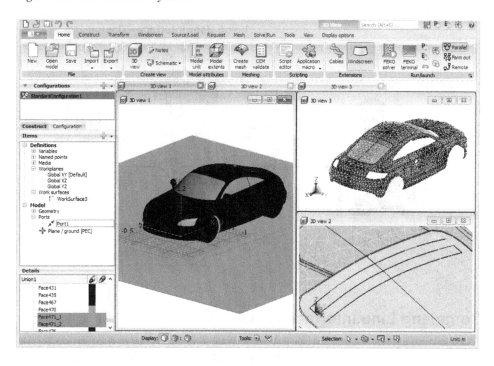

automation solutions allow enterprises to reach both these goals. However, end-users are becoming increasingly technology savvy, and thus speech-enabled applications must evolve quickly to meet consumer needs. Users expect highly effective, efficient solutions that are likable and quickly learned (Microsoft, 2015).

A Voice User Interface is what a person interacts with when communicating with a spoken language application. The elements of a VUI include prompts, grammars, and dialog logic (also referred to as call flow). The prompts, or system messages, are all the recordings or synthesized speech played to the user during the dialog. Grammars define the possible things callers can say in response to each prompt. The system can only understand those words, sentences, or phrases that are included in the grammar. The dialog logic defines the actions taken by the system, for example, responding to what the caller has just said or reading out information retrieved from a database (Cohen M.H, Giangola J.P and Balogh J, 2004).

Figure 16. Voice user interface

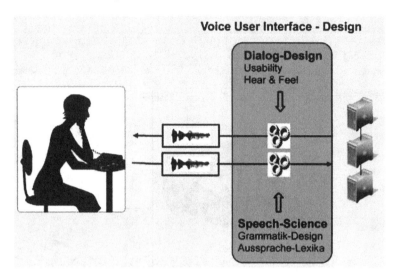

Command Line Interfaces (CLI)

Sometimes referred to as the command screen or a text interface, the command line is a user interface that is navigated by typing commands at prompts, instead of using the mouse (Computer Hope, 2015). Command line interface

Figure 17. Touch Interface application
Texas Instruments, 2014.

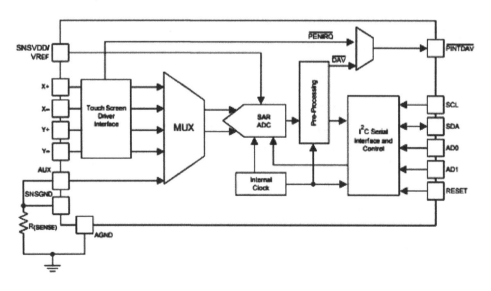

Figure 18. The plant layout proposed by the researcher in fuzzy with CBM

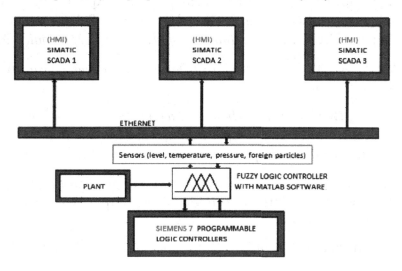

(CLI) is a text-based interface that is used to operate software and operating systems while allowing the user to respond to visual prompts by typing single commands into the interface and receiving a reply in the same way (Janalta Interactive Inc, 2010).

Touch Interfaces (TI)

Touch-screen interfaces are effective in many information appliances, in personal digital assistants (PDAs), and as generic pointing devices for instrumentation and control applications (Downs, 2005).

In this book, the researcher is going to focus on the GUI since PLCs are to be used with SCADA system to monitor the parameters of the plant.

This research shows a gap of fuzzy logic and condition based maintenance in trying to reduce machinery failure and Figure 18 is a schematic layout to be accomplished in the research.

REFERENCES

Advameg, Inc. (2015, October 12). *Automation*. Retrieved from Encyclopedia of Small Business: http://www.referenceforbusiness.com/small/A-Bo/Automation.html

Ahmadi, H., Moosavian, A., & Khazaee, M. (2012). An Appropriate Approach for Misalignment Fault Diagnosis Based on Feature Selection and Least Square Support Vector Machine. *International Journal of Mechanics*, 33-42.

Alhaddad, M. A. (2013). *Optimize wash time of washing machine using fuzzy logic*. Academic Press.

Atmel. (2015, December 15). *Human Machine Interface*. Retrieved from Atmel: http://www.atmel.com/applications/industrialautomation/human_machine_interface/default.aspx

Babuska and Mamdani. (2008, November 2). *Fuzzy control scholarpedia*. Retrieved September 27, 2013, from Fuzzy control: http://www.scholarpedia.org/article/Fuzzy_control

Berger. (2010). *Ten Pitfalls to Avoid When Selecting a CMMS/EAM*. CMMS.

Charwat, H. J. (1992). *Lexikon der Mensch-Maschine-Kommunikation*. Oldenbourg.

Cohen, M. H., Giangola, J. P., & Balogh, J. (2004). Voice User Interface Design. In *Introduction to Voice User Interfaces* (pp. 1-36). Addison Wesley. Retrieved December 16, 2015, from http://ldt.stanford.edu/~ejbailey/02_FALL/ED_147X/Readings/CohenExcerpt.Winograd.pdf

Dash. (2010). Fuzzy-logic based trend classification for fault diagnosis of chemical processes. *Computers & Chemical Engineering*, 347–362.

Downs. (2005). Using resistive touch screens for human/machine interface. *Analog Applications Journal*, 5-10.

Dunn. (1997). Implementing a Computerized Maintenance Management System, Why Most CMMS Implementations Fail to Provide the Promised Benefits. In *Maintenance in Mining Conference* (pp. 55-71). Sydney, Australia: Maintenance in Mining Conference.

Encyclopaedia. (2015, December 15). Retrieved from Definition of:HMI: http://www.pcmag.com/encyclopedia/term/44300/hmi

Guimaraes, A. C. F., & Lapa, C. M. F. (2007). Fuzzy inference to risk assessment on nuclear engineering systems. *Applied Soft Computing*, 7(1), 17–28. doi:10.1016/j.asoc.2005.06.002

Hope, C. (2015). *Dictionary*. Retrieved from Dictionary: http://www. computerhope.com/jargon/c/commandi.htm

Innovolt. (2014, August 18). Retrieved March 11, 2015, from Powering performance: http://innovolt.com/

Iserman. (1998). On fuzzy logic applications for automatic control supervision, and fault diagnosis. *IEEE Transaction Systems*, 221-235.

J, T. D. (2010). *USA Patent No. 7,859,519 B2*. Washington, DC: US Patent Office.

Jolandan, S. G., Mobli, H., Ahmadi, H., Omid, M., & Mohtasebi, S. S. (2012). Fuzzy-Rule-Based Faults Classification of Gearbox Tractor. *WSEAS Transactions on Applied and Theoretical Mechanics*, 50-62.

Khan, M. K., & Hafiz, N. (1999). Development of an expert system for implementation of ISO quality systems. *Total Quality Management, 10*(1), 47–59. doi:10.1080/0954412998054

Koditek F. (2012, March 13). How to improve energy management. *Manufacturing Automation*.

Konstandinidou, M., Nivolianitou, Z., Kiranoudis, C., & Markatos, N. (2006). A Fuzzy modeling application of CREAM methodology for human reliability analysis. *Reliability Engineering & System Safety, 91*(6), 706–716. doi:10.1016/j.ress.2005.06.002

Kumar, A., Sharma, S. P., & Kumar, D. (2007). *Robot reliability using petri nets and fuzzy lambda-tau methodology. In 3rd International Conference on Reliability and Safety Engineering* (pp. 28–37). Reliability and Safety Engineering.

Microsoft. (2015, December 16). *Voice User Interface Design - Purpose and Process*. Retrieved from Microsoft: https://msdn.microsoft.com/en-us/library/ms994650.aspx

Momeni, M., Fathi, M. R., Zarchi, M. K., & Azizollahi. (2011). A fuzzy TOPSIS based approach to maintenance strategy selection: A case study. *Middle East Journal of Scientific Research*, 699-706.

Mostafa, M., Payam, M., Hatami, A. M., Alireza, G., & Allah, K. H. (2013). A Study of Barriers and Success Keys to The Implementation of Computerized Maintenance Management System in an Organization: Case Study in Fan Avaran Petrochemical Company. *Life Science Journal*, 20–34.

Mustapha, F., Sapun, S. M., Ismail, N., & Mokhtar, A. (2004). A computer based intelligent system for fault diagnosis of an aircraft engine. *Engineering Computation*, *21*(1), 78–90. doi:10.1108/02644400410511855

Passino, K., & Yurkovich, S. (1998). *Fuzzy control*. Menlo Park, CA: Addison Wesley Longman, Inc.

Rzevski. (1995). Artificial Intelligence in Engineering: Past, Present and Future. *AIENG95 Conference* (pp. 31-40). Udine, Italy: AIENG.

Savoia. (2002). Structural reliability analysis through fuzzy number approach, with application to stability. *Computers & Structures*, 1087–1102.

Schneider. (1993). Implementation of a fuzzy concept for supervision and fault detection of robots. *First European Congress on Fuzzy and Intelligent Technologies* (pp. 775-780). London: Fuzzy Technologies.

Sharma, R. K., Kumar, D., & Kumar, P. (2007). Quality costing in process industries through QCAS: A practical case. *International Journal of Production Research*, *45*(15), 3381–3403. doi:10.1080/00207540600774067

Subnet Solutions Inc. (2015, December 15). *Human Machine Interface*. Retrieved from Dictionary: http://www.subnet.com/resources/dictionary/human-machine-interface.aspx

Tao, Chan, & Wang. (2013). An approach to performance assessment and fault diagnosis for rotating machinery equipment. EURASIP Journal on Advances in Signal Processing, 738-746.

Texas Instruments, . (2014). *TSC2013-Q1 12-Bit, Nanopower, 4-Wire Dual-Touch Screen Controller With I2R*. Texas Instruments.

Wang, W. L., Pan, D., & Chen, H. M. (2006). Architecture-based software reliability modeling. *Journal of Systems and Software*, *79*(1), 132–146. doi:10.1016/j.jss.2005.09.004

WebFinance, Inc. (2015, October 23). *Graphical user interface (GUI)*. Retrieved from Business Dictionary: http://www.businessdictionary.com/definition/graphical-user-interface-GUI.html#ixzz3uPQzxdIX

Wu By, C. S., Polte, T., & Rehfeldt, D. (2001). *A Fuzzy Logic System for Process Monitoring and Quality Evaluation in GMAW*. American Welding Society and the Welding Research Council.

Yager, R. R., & Filev, D. P. (1994). *Essentials of Fuzzy Modeling and Control*. John Wiley & Sons Ltd.

Yuan, Y. F., & Zhuang, H. J. (1996). A genetic algorithm for generating fuzzy classification rules. *Fuzzy Sets and Systems, 84*(1), 1–19. doi:10.1016/0165-0114(95)00302-9

Chapter 4
Methodological Approaches on Fuzzy Logic Applications in Selected Companies

ABSTRACT

The chapter is based on fuzzy logic approaches to machinery failure analysis. Six companies and organisations were used as model validation case studies. At a platinum mining company, the research was based on the root cause analysis technique. The objective was to determine the major causes of failure of the pebble crusher, to estimate between the major crusher failures, and to provide suitable solutions that included the optimization of the crushing circuit. The second application was done on beverages manufacturing companies focusing on the bottle washing process. The third company used a reactive and firefighting maintenance strategy which resulted in frequent catastrophic breakdowns, ever increasing maintenance costs, and long, unplanned plant shutdowns. The fourth was done for a hydropower generation company. The dynamic characteristics of these systems are nonlinear and difficult to predict. The fifth case study application focused on the water gate control and avoidance of its failure. The sixth and final case study application was at a thermal power generation company.

INTRODUCTION

Measuring knowledge management is a critical basis for developing incentives for further stimulating knowledge sharing and networking on local and global levels. Without quantifiability, measurement endeavours remains elusive.

DOI: 10.4018/978-1-5225-3244-6.ch004

Further, it is critical to ensure that existing knowledge assets are constantly challenged in a purposeful way. Especially in the current Internet Age, where today's core competencies quickly turn into tomorrow's core rigidities, it is incumbent upon companies to ensure that the knowledge they nurture inside is still relevant to the market thus, the need to explicitly address the issue of developing metrics and incentives for knowledge management. The key success factors of developing a working AI at a company are:

- Organisation
- Strategy, systems and infrastructure
- Effective and systematic processes
- Measures
- Costs
- Management support

METHODS OF THE STUDY

When conducting a research, one has to use either quantitative or qualitative method. The qualitative format is in the form of an essay whilst a quantitative makes use of numerical scoring and grading.

On qualitative research basis for the course of action (Bryman, Social Research Methods, 2004) (Mathi, 2004) and (Bryman, Quantity and Quality in Social Research, 1988)mentions that there are five ways to choose from, i.e

- Experiment,
- Survey,
- Qualitative research,
- Case study and
- Action research (Mathi, 2004)

According to (Alasuutari, 1996) a qualitative research process involves two phases: the Purification of Observations and the Unriddling.

The research would review all the stages from review up to the management level; what they can understand in simple terms.

Figure 1. System's development methodology
Mathi, 2004.

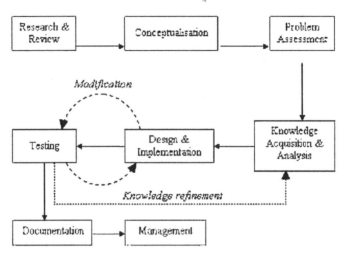

Quantitative vs. Qualitative

Quantitative research is based on the measurement of quantity or amount. It is applicable to phenomena that can be expressed in terms of quantity. Qualitative research, on the other hand, is concerned with qualitative phenomenon, i.e, phenomena relating to or involving quality or kind. For instance, when we are interested in investigating the reasons for human behaviour (i.e., why people think or do certain things), we quite often talk of 'Motivation Research', an important type of qualitative research, this type of research aims at discovering the underlying motives and desires, using in depth interviews for the purpose. Other techniques of such research are word association tests, sentence completion tests, story completion tests and similar other projective techniques.

Attitude or opinion research i.e., research designed to find out how people feel or what they think about a particular subject or institution is also qualitative research. Qualitative research is especially important in the behavioural sciences where the aim is to discover the underlying motives of human behaviour. Through such research we can analyse the various factors which motivate people to behave in a particular manner or which make people like or dislike a particular thing.

The above description of the types of research brings to light the fact that there are two basic approaches to research, viz., *quantitative approach* and the *qualitative approach*. The former involves the generation of data in quantitative form which can be subjected to rigorous quantitative analysis in a formal and rigid fashion. This approach can be further sub-classified into *inferential, experimental* and *simulation approaches* to research. The purpose of *inferential approach* to research is to form a data base from which to infer characteristics or relationships of population. This usually means survey research where a sample of population is studied (questioned or observed) to determine its characteristics, and it is then inferred that the population has the same characteristics.

Experimental approach is characterised by much greater control over the research environment

and in this case some variables are manipulated to observe their effect on other variables. *Simulation approach* involves the construction of an artificial environment within which relevant information and data can be generated. This permits an observation of the dynamic behaviour of a system (or its sub-system) under controlled conditions. The term 'simulation' in the context of business and social sciences applications refers to "the operation of a numerical model that represents the structure of a dynamic process.

Given the values of initial conditions, parameters and exogenous variables, a simulation is run to represent the behaviour of the process over time. Simulation approach can also be useful in building models for understanding future conditions. *Qualitative approach* to research is concerned with subjective assessment of attitudes, opinions and behaviour. Research in such a situation is a function of researcher's insights and impressions. Such an approach to research generates results either in non-quantitative form or in the form which are not subjected to rigorous quantitative analysis. Generally, the techniques of focus group interviews, projective techniques and depth interviews are used (Bryman, Quantity and Quality in Social Research, 1988).

THE SCIENCES OF THE ARTIFICIAL INTELLIGENCE

The engineer, and more generally the designer, is concerned with how things ought to be - how they ought to be in order to attain goals, and to function. With goals and "oughts" we also introduce into the picture the dichotomy between normative and descriptive. Natural science has found a way to exclude the normative and to concern itself solely with how things are done. Artificial things can be characterized in terms of functions, goals and adaptation.

Table 1. Research quadrant

	Theory Oriented	**Practice-Oriented**
Empirical	Physics; Sociology	Political poll
Design	Develop incremental structured parser	Develop website

Quantitative Approaches

Strengths

- Precision: through quantitative and reliable measurement
- Control: through sampling and design
- Ability to produce causality statements, through the use of controlled experiments
- Statistical techniques allow for sophisticated analysis
- Replicable

Limitations

- Because of the complexity of human experience it is difficult to rule out or control all the variables;
- Because of human agency people do not all respond in the same ways as inert matter in the physical sciences;
- Its mechanistic ethos tends to exclude notions of freedom, choice and moral responsibility;
- Quantification can become an end in itself.
- It fails to take account of people's unique ability to interpret their experiences, construct their own meanings and act on these.
- It leads to the assumption that facts are true and the same for all people all of the time.
- Quantitative research often produces banal and trivial findings of little consequence due to the restriction on and the controlling of variables.
- It is not totally objective because the researcher is subjectively involved in the very choice of a problem as worthy of investigation and in the interpretation of the results.

Qualitative Approaches

Limitations

- The problem of adequate validity or reliability is a major criticism. Because of the subjective nature of qualitative data and its origin in single contexts, it is difficult to apply conventional standards of reliability and validity.
- Contexts, situations, events, conditions and interactions cannot be replicated to any extent nor can generalisations be made to a wider context than the one studied with any confidence.
- The time required for data collection, analysis and interpretation is lengthy.
- Researcher's presence has a profound effect on the subjects of study.
- Issues of anonymity and confidentiality present problems when selecting findings.
- The viewpoints of both researcher and participants have to be identified and elucidated because of issues of bias.

Strengths

- Because of close researcher involvement, the researcher gains an insider's view of the field. This allows the researcher to find issues that are often missed (such as subtleties and complexities) by the scientific, more positivistic enquiries.
- Qualitative descriptions can play the important role of suggesting possible relationships, causes, effects and dynamic processes.
- Because statistics are not used, but rather qualitative research uses a more descriptive, narrative style, this research might be of particular benefit to the practitioner as she or he could turn to qualitative reports in order to examine forms of knowledge that might otherwise be unavailable, thereby gaining new insight.
- Qualitative research adds flesh and blood to social analysis.

Quantitative Research Design

The emphasis is on how each and every member at the five companies come up with the better ways of CBM using AI which is healthier for the plant to avoid minor stoppages within production times. The research will be entirely carried out in developing nations. Physical visits to all the companies was done analysing each type of maintenance and why opting for fuzzy-CBM.

TOOLS TO BE USED AND HOW TO USE THEM

- Ishikawa diagram using Edraw Max software (Finding the reason why a certain problem persists)
- Bar graphs (Arranging data and be understood faster and clearer)
- PDCA cycle (Reviewing the processes)
- Matlab software (Input data, model and simulate fuzzy logic systems)
- FMECA; Failure Mode Effects and Criticality Analysis (Root cause analysis of the problems)
- Decision fault trees (Grouping maintenance strategies)
- Auto CAD 2010

MATLAB due to the fact that it is modern and solves a lot of problems is going to be the centre of the analysis tool.

The Language of Technical Computing

Matlab is an abbreviation for "matrix laboratory". While other programming languages mostly work with numbers one at a time, Matlab is designed to operate primarily on whole matrices and arrays. All Matlab variables are multidimensional arrays, no matter what type of data. A matrix is a two-dimensional array often used for linear algebra. MATLAB is a high-level language and interactive environment that enables you to perform computationally intensive tasks faster than with traditional programming languages such as C, C++, and FORTRAN.

Key Features of Matlab

- High-level language for technical computing
- Development environment for managing code, files, and data
- Interactive tools for iterative exploration, design, and problem solving
- Mathematical functions for linear algebra, statistics, Fourier analysis, filtering, optimization, and numerical integration

- 2-D and 3-D graphics functions for visualizing data
- Tools for building custom graphical user interfaces
- Functions for integrating MATLAB based algorithms with external applications and languages, such as C, C++, FORTRAN, Java, COM, and Microsoft Excel.

QUALITY OF THE PROJECT

Condition monitoring is the process of monitoring the operating characteristics of a machine so that changes and trends of the monitored characteristics can be used to predict the need for maintenance before serious deterioration or breakdown occurs and to estimate the health of the machine. Condition Based Maintenance (CBM) is the process of planning and optimizing maintenance activities based on condition monitoring.

Recently, there have been considerable research efforts to develop condition monitoring technologies for industrial processes and systems. The use of these technologies has resulted in the acquisition of large amounts of data and has given life to new fields of research, namely data mining and knowledge discovery in databases. In the past, one of the main problems in maintenance management was the lack of data needed to support the decision making process. Nowadays, most companies use one or more condition monitoring technologies and possess considerable databases containing performance indicators for their machines. Consequently, researchers are now interested in finding techniques to extract information and interpret it accurately. The objective of the project is to use human machine interface and monitor the plant while in the office as in Figure 2.

DATA ANALYSIS

All the companies visited and researched on will be modelled to proper ways of maintenance to avoid machinery failure. Machinery with highest number of failures will be looked at and an optional solution is advised.

FUZZY INFERENCE SYSTEM

Steps in fuzzy inference system:

Figure 2. Human machine interface (double click to see in detail the excel document)

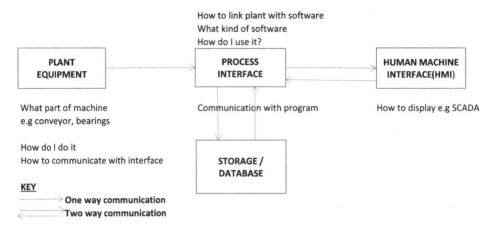

Step 1: Fuzzification

The first step in a fuzzy inference system is the fuzzification of crisp inputs. It transforms the exact logic problem into a fuzzy logic problem. Unlike crisp logic, fuzzy logic deals with linguistic variables instead of numerical variables. The process of converting numerical variables of the problem into grades of membership for linguistic terms of fuzzy sets is called fuzzification. Thus it is a mapping from a certain input space to fuzzy sets in certain input universes of discourse.

Step 2: Rule Evaluation

The next step in the fuzzy inference system is to apply the fuzzified inputs to the antecedents of the fuzzy rules. In case a given fuzzy rule has more than one antecedent, we make use of the fuzzy operator AND or OR in order to obtain a single truth value that would represents the result of the antecedent evaluation.

To evaluate the conjunction (intersection) & disjunction (union) of the rule antecedents, the fuzzy operators AND & OR are used respectively.

$$AND : \mu A \cap B(x) = \min\left[\mu A(x), \mu B(x)\right] \tag{1}$$

$$OR : \mu A \cup B(x) = \max\left[\mu A(x), \mu B(x)\right] \tag{2}$$

Figure 3. Operation of fuzzy systems

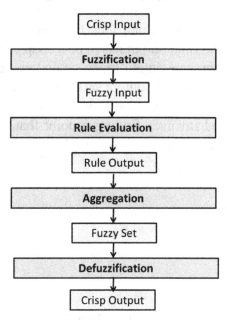

Then the result of the evaluation would be applied to the consequent membership function. There are two main methods of doing so:

1. **Clipping:** This involves cutting (alpha-cut) the consequent membership function at the level of result of the antecedent evaluation. As the top of the membership function is sliced, some information loss is inevitable in case of clipping. Still it is often preferred as it doesn't involve too complex mathematics.
2. **Scaling:** In this case, the membership functions of the rule consequent are adjusted by multiplying all its membership degrees by the truth value of the rule antecedent. There is not much loss of information.

Step 3: Aggregation of the Rule Output

This is the process of uniting the outputs of all rules that are invoked for a particular set of inputs into a single fuzzy set. The clipped or scaled consequent membership functions serve as the input to the aggregation process and the output of the process is one fuzzy set for each output variable.

Step 4: Defuzzification

It is the final step in the fuzzy inference process. It is the reverse process of Fuzzification. The Fuzzy Logic Controller (FLC) produces required output in a fuzzy set; however the final output has to be a crisp value. Defuzzification involves taking the aggregate output fuzzy set and producing a single crisp value corresponding to each of the output variables. Of various defuzzification methods, the centroid defuzzification method is most commonly used. Centroid defuzzification method involves finding a point that would represent the Centre of gravity of the aggregate fuzzy set. Mathematically this point can be expressed as:

$$COG = \frac{\int_a^b \mu_A(x)x\,dx}{\int_a^b \mu_A(x)\,dx} \qquad (3)$$

Importing and Exporting From the GUI Tools

When you save a fuzzy system to a file, you are saving an ASCII text FIS file representation of that system with the file suffix .fis. This text file can be edited and modified and is simple to understand. When you save your fuzzy system to the MATLAB workspace, you are creating a variable (whose name you choose) that acts as a MATLAB structure for the FIS system. FIS files and FIS structures represent the same system.

ANALYSIS OF FUZZY LOGIC TECHNIQUE

Fuzzy logic is a machine intelligent technique which can possibly be implemented in optimizing the CBM process. FL offers several unique features that make it a particularly good choice for many control problems including among other features the following:

1. It is inherently robust since it does not require precise, noise-free inputs and can be programmed to fail safely if a feedback sensor quits or is

destroyed. The output control is a smooth control function despite a wide range of input variations.

2. Since the FL controller processes user-defined rules governing the target control system, it can be modified and tweaked easily to improve or drastically alter system performance. New sensors can easily be incorporated into the system simply by generating appropriate governing rules.

3. FL is not limited to a few feedback inputs and one or two control outputs, nor is it necessary to measure or compute rate-of-change parameters in order for it to be implemented. Any sensor data that provides some indication of a system's actions and reactions is sufficient. This allows the sensors to be inexpensive and imprecise thus keeping the overall system cost and complexity low.

4. Because of the rule-based operation, any reasonable number of inputs can be processed (1-8 or more) and numerous outputs (1-4 or more) generated, although defining the rule base quickly becomes complex if too many inputs and outputs are chosen for a single implementation since rules defining their interrelations must also be defined. It would be better to break the control system into smaller chunks and use several smaller FL controllers distributed on the system, each with more limited responsibilities.

5. FL can control nonlinear systems that would be difficult or impossible to model mathematically. This opens doors for control systems that would normally be deemed unfeasible for automation.

CONCLUSION

Condition monitoring is the process of monitoring the operating characteristics of a machine so that changes and trends of the monitored characteristics can be used to predict the need for maintenance before serious deterioration or breakdown occurs and to estimate the health of the machine. Condition Based Maintenance (CBM) is the process of planning and optimizing maintenance activities based on condition monitoring. AI has been very successful in limited problem domains especially complicated like for the example at Beverages Manufacturing Company (BMC), Platinum Mining Company (PMC), Fruits Processing Company (FPC), Hydropower Generation Company (HPDC), Thermal Power Generation Company (TPGC) and Water Distribution Company (WDC) it can work because the plants are complex, automobile plants also have got complicated robots that need intelligent control and monitoring. This can also be called intelligent manufacturing. Intelligent manufacturing is the use of production process technology that can automatically adapt to changing environments and varying process requirements, with the capability of manufacturing various products with minimal supervision and assistance from operators. The development and implementation of artificial intelligence in manufacturing is growing rapidly. Qualitative analysis is going to be carried out to collect data. The companies were selected because of different reasons: BMC is the only company in the country selected with state of the art robots and complex machinery, PMC produces platinum and breakdown of equipment is a disaster leading to no production at all. FPC also bought the robots that control the whole process of production. HPDC is the only power plant that uses hydro mechanism and on its own is the eye of the nation to produce power. TPDC deals with coal based power generation and machinery is important to be monitored to avoid failure and lastly WDC is the city council company that operates the movement of water.

REFERENCES

Alasuutari. (1996). Theorizing in Qualitative Research: A cultural Studies Perspective, University of Tampere, Finland. *Qualitative Inquiry*, 20–29.

Bryman. (1988). *Quantity and Quality in Social Research.* London: Routledge.

Bryman. (2004). *Social Research Methods.* Oxford, UK: Oxford University Press.

Mathi. (2004). *Key Success Factors for Knowledge Management.* University of Applied Sciences/FH Kempten.

KEY TERMS AND DEFINITIONS

BMC: Beverages manufacturing company.
FPC: Fruits processing company.
HPDC: Hydropower generation company.
PMC: Platinum mining company.
TPGC: Thermal power generation company.
WDC: Water distribution company.

Chapter 5

Fuzzy Logic Application in Improving Maintenance in a Beverage Manufacturing Company

ABSTRACT

Fuzzy logic approach was done on a beverage manufacturing company focusing on the bottle washing process. The main problem is that the pneumatic valve of the bottle washer that controls the discharge of clean bottles sometimes sticks or fails, which results in significant loss of production since this is a bottleneck operation. The main causes of failure were found to be the temperature and pressure, which often fell outside the required ranges, and minor contributions to failure due to moisture and abrasive particles. In order to solve this problem, a model reference adaptive fuzzy controller was designed for the pneumatic valve using the MATLAB software. The model reference adaptive control (MRAC) system consists of the reference model that has the desired output of the system. The error resulting from the difference between the actual system output and that of the reference model is executed by the fuzzy logic controller (FLC). The simulation of the behaviour of the valve in response to the reference model was done using Simulink.

DOI: 10.4018/978-1-5225-3244-6.ch005

NEW HIGHLY AUTOMATED PRODUCTION PLANT INSTALLED IN DEVELOPING NATION OF AFRICA

BMC has installed a state of the art Mechatronics controlled filling system. The new plant was manufactured in Germany by the company called Krones. The Krones Group, headquartered in Neutraubling, Germany, plans, develops, and manufactures machines and complete lines for the fields of process technology, bottling, canning, packaging and intralogistics (Krones, 2012).

Every day, millions of bottles, cans and specially-shaped containers are "processed" on lines from Krones; particularly in breweries, the soft-drinks sector and at still-wine, sparkling-wine and spirits producers, but also in the food and luxury goods sectors, as well as the chemical, pharmaceutical and cosmetic industries. Since being founded in 1951, Krones has evolved far beyond its original role as a mere producer of machinery and bottling lines. The company has meanwhile become an "all-round partner" for its customers, creating harmonious, optimised synergies of mechanical engineering, line-related expertise, process technology, microbiology and information technology. Today, Krones is synonymous with "systems engineering" (Krones, 2012).

Line 1 at BMC

The design for an assembly / production line is determined by analysing the steps necessary to manufacture each product component as well as the final product. All movement of material is simplified, with no cross flow, backtracking, or repetitious procedure. Work assignments, numbers of machines, and production rates are programmed so that all operations along the line are compatible (Encyclopedia, 2011). Line 1 at BMC starts from the palletizer where empty dirty/clean bottles enter into production up to the depalletizer whereby filled bottles with product are ready for the market.

The data shows that the bottle washer is the one that has a lot of break downs, 2209 minutes (36.816700 hours) in January 2012 only, Table 2 and Figure 1. It is then followed by the palletizer. Bottle conveyors, empty bottle inspector and the filler are also having some breakdowns.

Table 1. Line 1 machine identification

	Machine Name	Line 1 Component	Machine Number
1	Mecafill	Filler	K129-B16
2	Lavatec	Bottle washer	K682-330
3	Modulpal	Palletiser	KR63-404
4	Modulpal	Depalletiser	KR63-405
5	Smartpac	Packer (Crater)	KR66-144
6	Smartpac	Unpacker (Uncrater)	KR66-143
7	Checkmat	Fulls Bottle Inspector (FBI)	K731-K21
8	Crate Checkmat	Fulls Crate Inspector	K708-771
9	Linatronic	ASEBI (Empty bottle inspector)	K735-765
10	Contiflow	Blender/Mixer	KB40-308
11	MultiCo	Case conveyors	KR67-899
12	PalCo	Pallet conveyors	KR57-989
13	Bottle Conveyor	Bottle Conveyor	K995-R7H
14	Diajet	Crate washer	K693-038
15	Case turner	Case turner	K691-104
16	VarioClean	CIP tanks	KB80-332
17		Decapper	
18		Capper	
19	Videojet	Date corder	

FILLER

The filler is where filling of the product occurs (Figure 2). Maintenance is usually done using preventive action.

The filler is a complex system in nature and needs attention when working with it to avoid unnecessary breakdowns. It contains Crowner inside; this has got some problems for jams and bursting of bottles. The filler injects the fluid into the bottle. Filling is primarily carried out through the use of gravity although vacuum fillers and pressurised fillers are also available. The filling can be carried out through 'level fill', 'volumetric fill' (mass flow rate) or 'weigh filling' (Brassington, 2009). Due to the short time taken to fill each container, extreme pressures are present especially when filling carbonated liquids. When filling, the bottle is raised to the filling nozzle where a tight seal is made and the flow of liquid can then begin.

Table 2. January 2012 machine downtimes

Machine	Downtime (min) January 2012
Bottle Washer	2209
Palletiser	955
Bottle Conveyors	663
ASEBI	596
Filler	462
Packer	389
Date Coder	208
Case conveyors	163
Depalletiser	107
Crate Washer	65
Unpacker	65
pallet magazine	58
Fulls crate checkmat	10
Fulls Bottle checkmat	0
Contiflow	0
Empties sighting Station	0
Decapper	0

NB: Bottle washer is the one that shows more problems and has the highest number of downtimes. This is in descending order in which downtimes occur to (i.) Bottle washer, (ii.) Palletizer, (iii.) Bottle conveyors, (iv.) Empty Bottle Inspector and the (v.) Filler.

Figure 1. January 2012 Pareto chart for downtimes

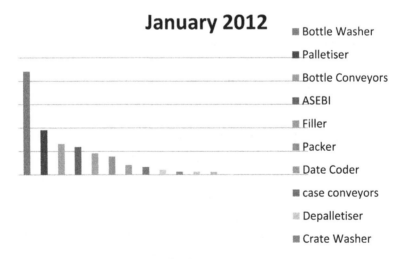

*For a more accurate representation see the electronic version.

Figure 2. Filler
Krones, 2012.

The hazards associated with filling machines include mechanical crushing and stabbing hazards between the container and the filling nozzle. High-pressure fluid is also a hazard particularly if the seal between the bottle and the filling nozzle is not tight. An impact hazard is present if a container shatters or when broken containers are ejected from the filling machine especially when carbonated fluids are being filled. There is also a hazard if flammable products are filled in a potentially explosive atmosphere (BSEN415-2:2000,

Figure 3. Bottle filler
Krones, 2012.

2015). It is obvious that carbonated liquids are considered far more hazardous than still liquids as can be seen from the extensive guarding on the carbonated liquid filling machine shown on the right. The still liquid filling machine is not so heavily enclosed. °C

This is a complex machine and is like the centre of all the processes, it has to use Heuristics and fuzzy logic to give more solutions to solve breakdown problems.

Problems for Filler

- Gas leakage from the distributor
- Wearing out of worms
- Bottle guides wear

Figure 4. Worn out worm inside the filler

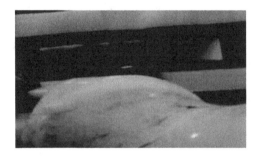

Figure 5. Servo drives in the filler
Krones, 2012.

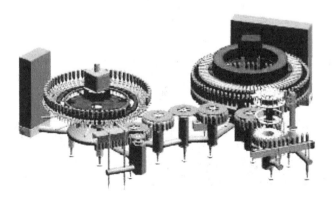

BOTTLE WASHER

Bottle washer has got some prepared sheets for PM in lubrication and maintenance strategy. Figure 6 shows the bottle washer currently working at BMC.

The bottle washer will have some problems of corrosion if water that comes from the boiler enters without recommended temperatures and pressures. From Figure 7, bottles move along a conveyor and enter into the bottle washer which cleans all the bottles thoroughly. Automation is applied a lot within this stage. Figure 7 shows the cleaned bottles on way to the filler.

A cleaning machine is used to ensure that any bottles being filled are free from contaminants. Bottles are washed both inside and out and then dried. The plan for a typical bottle washing machine is shown in figure above.

Figure 6. Bottle washer in action
Krones, 2012.

Figure 7. Bottle washer side view
Krones, 2012.

Hazards associated with cleaning machines include mechanical crushing and stabbing between cleaning nozzle and container (BS EN 415-2:2000). The temperature and chemical composition of the cleaning fluid is also a hazard as it is typically hot and toxic (Krones, 2012).

Figure 8. Washer breakdown analysis

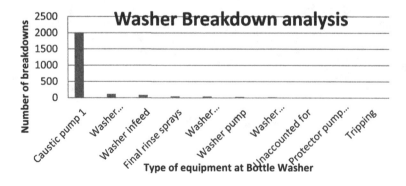

Intelligent Monitoring Method

The parameters in this machine can be accomplished in monitoring by using the fuzzy logic. Temperature variations that comes from the boiler might cause some problems to the bottle washer, so poka-yoke devices can be used to control how hot and cold the water is before entering the machine.

What to Monitor at the Washer

It is also of particular interest to monitor the water that enters the bottle washer to avoid corrosion in the elements of the washer. Figure 9 shows what can be monitored and controlled to cater for the prevention of rust and wear due to foreign particles (USPolicy, 2013).

Requirements for Cleaning Bottles

- Free of beverage altering Micro-organisms
- Free of any remains on the bottle (labels, tinfoil etc.)
- Free of any remains in the bottle (residual Liquids, etc.)
- Free of any odors

Table 3. Technical data for the bottle washer

Technical Data: Bottle Washer at BMC	Volume of Liquids
Capacity 50,400bph	High pressure prejetting 0.6m³
Min adjustment range 21,000bph	Presoak 2.58m³
Max adjustment range 61,000bph	Prespraying 1 and 2; 0.30m³
Processing time 24.14mins	Caustic Tank1; 57.10m³
Cycle time 2.93mins	Caustic Tank2; 57.10m³
Bottles per carrier 41	Caustic Tank3; N/A
No of carriers 615	Post caustic 7.77m³
Number of bottles per machine 20,295PCE	Warm water 1; 2.26m³
Empty weight 135t	Warm water 2; 2.56m³
Operating weight 279t	Cold water; 2.17m³
Machine length 17,053m	**Consumptions:**
Machine height 5.2m	Total water consumption 11.38m³/h
Machine width 9,515m	FW cons/bottle (0.750l) 0.23l/bottle

Krones, 2012.

Figure 9. Components to be monitored at the washer
Institute for 21ˢᵗ Century Energy, 2012.

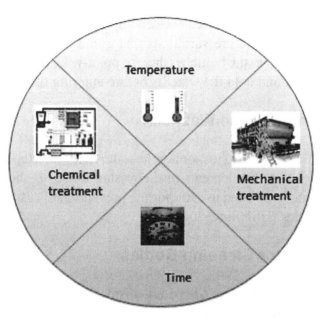

- Free of any chemical remains
- Brightness of the glass
- Free of droplets
- Adapted discharge temperature for further processing (Krones, 2012).

The broad range of Krones bottle washers provides successful cleaning for all applications: In sophisticated concepts, the glass and plastic bottles, which are in use worldwide since many years, will be extensively cleaned and gently treated. Clean and highly polished beverage and food packs meet the marketing requirements for an optimal presence of your product in the market. The machines of the single-end series KE are compact and very efficient, and they can be tailored to fit any individual application. The extensive variety includes different guard heights and offers a broad range for combining different modules for the heating and caustic zones. The chain loop guide inside the machines allows for a long retention period of the bottles in the baths. Furthermore, the arrangement of the label removal is optimally adjusted to the line performance. Another plus point of the machine features is the gentle container conveyance especially in the container in feed and discharge area.

The following problems make the bottle washer inefficient. One of them might be a cause that stops the washer and increase downtimes for the equipment. The information was obtained from the engineers (two of them), artisans (three) and operators (three) and also from the records of maintenance history since the plant was installed in 2011.

- Chain adjustment problems
- Pneumatic valve sticks sometimes (may give a sign that it is closed whilst it is not or vice-versa)
- Gearbox jamming

CONVEYOR

Figure 10 shows the conveyors at BMC. Items are moving along the conveyors.

There are quite a number of conveyors in the plant and are much wearing in the handling parts of occurs.

Despite the benefits of an effective maintenance engineering function, many companies fail to make it a top priority – and suffer the results; unreliable

Table 4. Consumption of the bottle washer

Consumption	Fresh Water	Condensate
Total consumption	8786 m³	1087MW/hr
Consumption per hour	5.4 m³/hr	333KW/hr
Consumption per bottle	110ml/Fl	
Consumption per bottle set point value	210ml/Fl	
Actual pressure	1.4 – 5 bars	
Minimum pressure	1.0 bar	
Steam pressure (Actual)	3.1bars	

Table 5. Dosing in the tanks

	Caustic 1 Tank	Caustic 2 Tank	Post - Caustic
Actual value	110mS	111mS	17mS
Set point value	110mS	110mS	15mS
Run-time of active ingredient in preparation	150Min.	150Min.	
Re-dosing active ingredient, sequential control.	20Sek	25Sek	

Table 6. Temperatures in the caustic tank

	Caustic 1 Tank	Caustic 2 Tank
Actual temperature ($^{\circ}$C)	77.1°C	76.3°C
Set Point ($^{\circ}$C)	77.0°C	77.0°C
	Condensate caustic 1	After counter current heater
Actual temperature ($^{\circ}$C)	77.0°C	85.5°C

Figure 10. Conveyors in action
Krones, 2012.

plant and equipment, reduced throughput, poor recoveries, excessive costs, and lost business. Maintenance Engineering is the discipline and profession of applying engineering concepts to the optimization of equipment, procedures, and departmental budgets to achieve better maintainability, reliability and availability of equipment. The objective of condition based maintenance (CBM) is typically to determine an optimal maintenance policy to minimize the overall maintenance cost based on condition monitoring information. Condition-based maintenance is imperative when critical components are concerned and their failure has sustained negative effects (on safety, environment, production, etc.) and/or when wear and ageing are involved. Line 1 is composed of the processes from mixture of ingredients up to when drinks are now ready for market. A typical bottling line performs the cleaning, filling, labelling and packing of bottles whether they are 'new' single-use bottles or reusable bottles. Figure below shows an example plan of a bottling line produced by Krones but in general the machines involved in a bottling line include a bottle unpacker, unscrambler and decapper, individual bottle washer/rinser/dryer, empty bottle inspection machine, filling machine, capper, filled bottle inspection machine, labelling and marking machines and bulk packing machines. Each of these machines has associated hazards. There are also some general hazards such as electrical, thermal, vibration and radiation hazards that should be considered for all machinery. Some failure to perform well must also be discussed and known.

Bottle Conveyors

Conveyors are used:

- When material is to be moved frequently between specific points
- To move materials over a fixed path
- When there is a sufficient flow volume to justify the fixed conveyor investment

Conveyors can be classified in different ways:

- Type of product being handled: *unit* load or *bulk* load
- Location of the conveyor: *overhead*, *on-floor*, or *in-floor*
- Whether or not loads can *accumulate* on the conveyor

At Delta there are two major types of conveyors which are the flexible conveyor and the automated roller conveyor. These are shown in figs below respectively. A flexible conveyor is used extensively in shipping/receiving operations for package handling, flexible conveyor is usually anchored at one end to fixed gravity or automated conveyor allowing the other end to be expanded and flexed into trailers for loading and unloading (Perry and Green, 1984). The automated version of gravity roller conveyor, automated roller conveyor is used extensively in large conveyor systems. A version of automated roller conveyor called Zero-Pressure Accumulating Conveyor is especially useful in avoiding the pressure build-up which normally occurs when product accumulates at a stationary operation. In KRONES plants, sensor signals are transmitted at moving conveyor elements using AS-Interface (Actuator Sensor Interface) with sliding contacts.

Problems

- Lack of lubrication on gearboxes and belts
- Slipping out of conveyor chain
- Slacking of chains

PALLETIZER

Figure 11 shows some problems and downtimes of the palletiser due to some components in the palletiser that were down by January 2012.

Palletizer needs much more attention for its maintenance because in Figure above it shows that some components break down up to a point when they are malfunctioning. Intelligent monitoring will be a good idea for this plant.

Palletizer Machine

The palletizer machine is a device that is used to stack factory goods and products on to a pallet. The first palletizer was built and installed by a company in 1948. Row-forming palletizers were introduced in the early 1950s that were used to arrange the goods in the row form. The in line palletizers were built in 1970s when high speed was needed for palletizing. In the early 1980s robotic palletizers machines were introduced which have an arm. A robotic palletizer machine is a complex machine that is used to pack boxes at a very high speed (RockwellAutomation, 2012)). At BMC they use robotic palletizers.

Figure 11. Palletizer downtimes

Palletiser	
nature of stoppage	down time
head crashing on pallet station	492
head dropping cases	366
table conveyor failure	44
Palletiser cylinder air connector	35
TOTAL	937

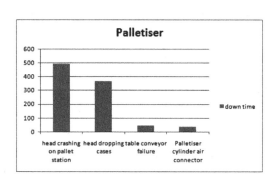

Table 7. Technical data for the palletizer

Technical Data	Robot Palletizers	Standard Palletizers
Number of axes	up to 6	
Loading capacity	350kg	180 kg / pattern large
Controls	KUKA, ABB	Allen Bradley, ELAU,
Production Lines	up to 4	up to 2
Additional Handling	slip sheets, empty pallets	slip sheets

Technical Information Concerning Palletizer

Palletizer other information

- **Bottles:** For 1.5L
- **Produce Capacity:** 1400 packs / min
- **Voltage:** 400 V / 50Hz
- **Voltage of Control:** 24V
- **Power:** 15kW + 2 x 0.55Kw conveyors
- **Pressure of Control:** 6 bars

Problems

- Robot malfunction
- High impact on putting down trays
- Cables loosely connected
- PLC controller (Siemens and Allen Bradley)

Figure 12. Palletizer grouping module
Krones, 2012.

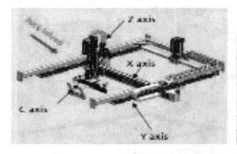

EMPTY BOTTLE INSPECTOR

Intelligent empty bottle inspection system is an indispensable inspection equipment of empty bottle before filling drinks. It is a blend of machine vision, precision machine and real-time control. Machine vision is the hard core, which affect the speed and accuracy of the inspection. Because bottles probably have some defects that may cause negative even dangerous consequences for production, glass bottles need to be checked before the products are canned in production. In many cases, this kind of work is performed manually. But manual inspection not only increases labor cost but is very difficult to guarantee inspection quality.

Machine Vision

This uses intelligence for the analysis of the bottles and makes a decision thereafter on rejecting or not depending on the nature of the bottle.

Quality control is an essential part of any bottle production system. When the bottles are produced in a manufacturing plant, they are inspected for manufacturing faults, However when empty bottles returned by the consumers are to be reused, the focus of quality control changes. Reusable bottles are

Figure 13. Robobox shutter head
Krones, 2012.

Shutter head

becoming increasingly popular due to their cost effectiveness. Inspection of the returned bottles, for re-using them, is required to ensure quality (Shafait F, Imran S M and Klette-Matzat S, 2004). This is as shown in Figures 14 and 15.

This is the one similar to that which is at BMC. Fuzzy logic is going to be used in this EBI to ensure CBM is carried out intelligently.

Worked same as Fulls Bottle Inspector; same working principle

Problems for EBI

• Sensors and cameras detect false information sometimes

If line 1 in the production facility is down for an hour it can cost millions of dollars in lost "opportunity revenue." The manufacturing process and maintenance need to be more complex and easy to solve problems. To satisfy customers and remain in business in this current world, intelligent

Figure 14. Machine Vision Inspector
Shafait F, Imran S M and Klette-Matzat S, 2004.

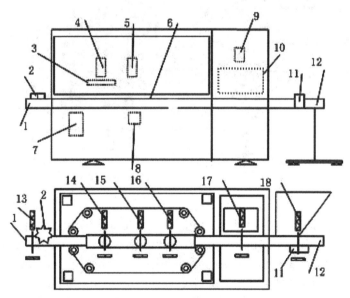

1-Bottle entry 2-Separator 3,8,10-Light 4,5,9-Camera
6-Conveyor 7-Cleaner 11-Rejector 12-Bottle output
13,14,15,16,17,18-Photoelectric sensor

Figure 15. Prototype of inspection
Liu H.J and Wang Y.N, 2008.

Figure 16. Layout of a fully labelled EBI
Krones, 2012.

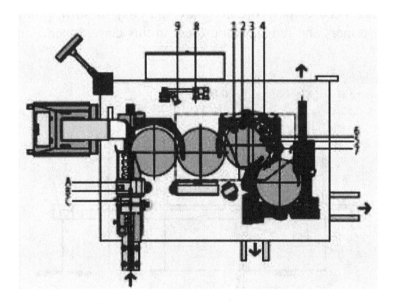

maintenance with fuzzy logic is one of the tools to boost the machinery and avoid downtimes. Much focus is to be given to the bottle washer since it is the one that is having more downtimes along line 1. Manufacturing intelligence is a strategy to collect critical, real-time data and turn it into insightful information that's visible and useful to people at every level of the organization. Manufacturing intelligence tools reduce costs and increase productivity by allowing employees to make informed decisions faster than ever before. Inevitably, that will make line 1 always being inspected in all its processes. Downtime is one of the most costly conditions a manufacturer can

Figure 17. Overview of an EBI
Krones, 2012.

experience; a proactive technical support program can generate significant cost savings. Continuous-monitoring services can also generate cost savings by protecting existing investments.

RESULTS AND ANALYSIS OF MAINTENANCE FUZZY LOGIC SIMULATION IN A BEVERAGES MANUFACTURING COMPANY

The authors went to BMC with the questionnaire and found the information they were looking for.

Results for BMC Line 1

From the research questionnaire designed by the researcher, artisans and one engineer, it reflected the following results. The titles of the various artisans and engineers in the company interviewed were as follows (twelve members): maintenance planner, machine specialist, asset care specialist, maintenance controller, packaging engineer, plant manager, electricians (x2), electrical foreman, operators and maintenance fitters (x2). The results are as shown in the Ishikawa diagrams

Ishikawa Diagrams for Bottle Washer

Ishikawa diagrams (also called fishbone diagrams, or herringbone diagrams, cause-and-effect diagrams, or Fishikawa) are causal diagrams that show the causes of a specific event -- created by Kaoru Ishikawa (1968). Common uses of the Ishikawa diagram are product design and quality defect prevention, to identify potential factors causing an overall effect. Each cause or reason for imperfection is a source of variation. Causes are usually grouped into major categories to identify these sources of variation. The categories typically include:

- **People:** Anyone involved with the process
- **Methods:** How the process is performed and the specific requirements for doing it, such as policies, procedures, rules, regulations and laws
- **Machines:** Any equipment, computers, tools etc. required to accomplish the job
- **Materials:** Raw materials, parts, pens, paper, etc. used to produce the final product
- **Measurements:** Data generated from the process that are used to evaluate its quality
- **Environment:** The conditions, such as location, time, temperature, and culture in which the process operates, Ishikawa 1968.

Figure 18. Ishikawa diagram of chain adjustment, drawn by Edraw Max

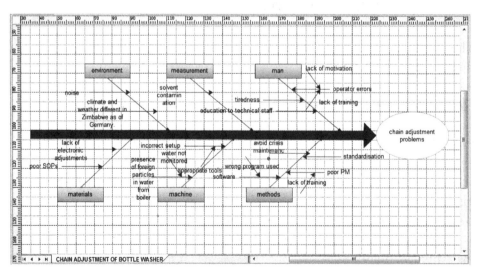

The pneumatic valve often sticks at start-up. Slack chain will cause uneven entry of teeth through the gears and overload occurs.

Too much torque shows too much messages of warning on control panel cause jamming.

Figure 19. Ishikawa diagram for pneumatic valve sticking, drawn by Edraw Max

Figure 20. Ishikawa diagram for gearbox jamming, drawn by Edraw Max

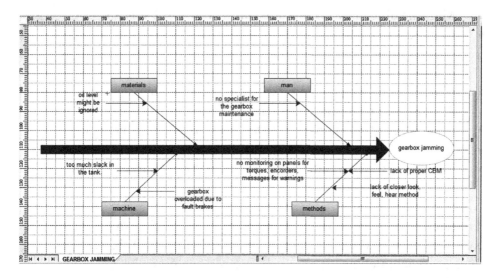

Failure Modes Effects and Criticality Analysis (FMECA)

Failure Modes Effects and Criticality Analysis (FMECA) is a quality tool which builds on the results of Functional Analysis to identify risks and their consequences. It grew out of Reliability Engineering efforts in the late 1950's. FMECA can be applied to systems, products, manufacturing processes, equipment, plant and even less tangible subjects such as logistic or information flows. It is used to identify the possible ways in which failure can occur the corresponding causes of failure, the corresponding effects of failure, and the impact on Customer Satisfaction. The objective of FMECA is to identify the components of products and systems most likely to cause failure, so that these potential failures can then be designed out. FMECA allows the identification early in the product development process of potential problems or safety hazards which are inherent in a product design. The safety and/or reliability of the product is assessed and modifications initiated at a relatively low cost before they are built into the product. Product reliability and customer satisfaction will be improved by preventing failures from occurring.

FMECA allows a Concurrent Engineering team to address reliability issues early in the design cycle where modifications are less costly, and critical risks associated with design or process concepts can be identified and the necessary corrective actions taken in time, Braglia 2000

Extra Data on the Bottle Washer From the BMC Team

- Machine efficiency for line 1 as on 27 May 2012 was 85%.
- Required water temperature from boiler to the bottle washer should range from 77°C to 95°C is accepted and will not harm anything; table below.
- Required pressure from boiler to the bottle washer should range from 3.5 bars to 5 bars is accepted and will not harm anything.
- Major type of maintenance at Delta is scheduled maintenance.
- Plant availability was 77%
- Coswin 7i is assisting in reducing breakdowns but is overpowered by breakdowns.

It is useful if the author do his intelligence to assist BMC in maintenance. The book will make BMC benchmark its practices to the world class performance since artificial intelligence responds faster to any error that occurs.

RESULTS OF FUZZY LOGIC INCORPORATED WITH MRAC TO MONITOR THE PLANT AT BMC

When failure occurs in any manufacturing process it is critical to identify and analyse the root causes leading to that failure. This helps to solve the real problem and to prevent recurrence. This chapter gives the possible causes of failure of the pneumatic valve of the Krones Bottle Washing Machine using the Ishikawa diagram. A Mamdani fuzzy controller is also proposed in this chapter to model the error changes in the system due to its ability to capture expert knowledge, its intuitive nature which allows it to be used for decision support applications and it has a widespread acceptance. The parameters to be monitored in this design are temperature, pressure, and dust, tilt angle of valve and moisture regulation. Two possible simulations are were to be carried out in this chapter and these include, the effect of temperature and pressure on the pneumatic valve using triangular membership function and the effect of dust and moisture on the valve using triangular membership functions. Simulation results of the system behaviour are going to be presented and analysed in this chapter. The MRAC system for the valve was also developed using the PID controllers and FLC. The results were simulated using Simulink.

Ishikawa Diagram of the Pneumatic Valve

The Ishikawa or cause and effect diagrams are commonly used for product design and prevention of quality defects so as to identify the root causes of the problem causing the failure of the pneumatic valve which in turn is causing the breakdown of the bottle washer. The sources of variation can be identified by grouping the causes into major categories. These categories include:

- **Environment:** These are the environmental conditions such as the temperature, culture, pressures and time in which the valve is operated.
- **Machines:** The tools required to complete the job, the computers and equipment are also analysed in order to determine their effect on the failure of the machine.
- **People:** The people involved in the process can contribute to the failure of the machine.
- **Materials:** This includes the raw materials required to accomplish the task, parts, paper and pen.

- **Methods:** This includes all the processes performed and the requirements for performing the process for example, rules, policies, regulations, procedures and laws.
- **Measurements:** This includes the data that is produced from the process in order to evaluate its quality.

Moisture and Dust Control

Effects of Moisture on the Pneumatic Valve

Moisture inside the valves may result from the compressed air passing through the valve. The effects of moisture on the valves include:

- Moisture may cause rusting in the moving parts of the valve.
- Increased rate of wear of the valve material may also result due to the moisture. The moisture washes the lubrication away which will result in the eventual failure or malfunctioning of the valve.
- The industrial processes which rely on the full functionality of the pneumatic control valves may be jeopardized and this usually results in costly breakdowns of the machine.

Figure 21. Ishikawa diagram for root cause analysis of pneumatic valve failure

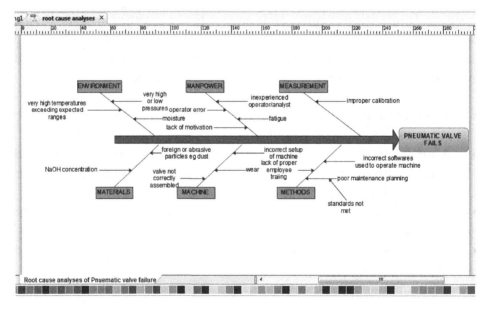

- Air or gas operated instruments may give inaccurate readings due to corrosion of the material and hence, interrupting the plant processes.
- The rubber diaphragms inside the pneumatic valves can be stiffened and will eventually rupture due to the moisture flowing through them.

The following are the range of values of moisture which are tolerated and some not tolerated inside the valve:
For a range of 1-5%

- **0-1%:** This range from 0-1% is the minimum amount of moisture that can enter the valve and causes no damage or less damage to the valve which will not result in the malfunctioning of the valve.
- **1-3%:** For these range of values of moisture entering the valve, the valve may still work efficiently opening up to 90° but further exposure to the moisture may result in the malfunctioning in the valve.
- **3-5%:** These ranges of moisture are not accepted inside the valve. These may result in wear, rusting and breakdown of the machine (Mushiri T., 2012).

Effects of Dust on the Pneumatic Valve

Foreign and abrasive particles causes scoring of the material and increases the rate of wear. Increased rate of wear results in reduced life span of the valves leading to failure. In order to prevent failure of the valves due to dust, a control system is going to be designed that will prevent unacceptable amounts of dust to enter through the valve.

The following are the range of values of dust accepted and not accepted inside the pneumatic valve:
For a range of 0-1%

- **0-0.02%:** This range from 0-0.02% is the minimum amount of dust that can enter the valve. For this range, the valve may still function properly without malfunctioning.
- **0.02-0.5%:** For 0.02-0.5% range of values of dust entering the valve, the valve may still work efficiently and open to the maximum angle up to 90° C. Further exposure to the dust may result in the malfunctioning in the valve.

- **0.5-1%:** These ranges of dust are not accepted inside the valve. These may result in wear, scoring and eventually breakdown of the machine (Mushiri T., 2012)

Development of Fuzzy Logic Rule Base for Moisture and Dust Control

The rule base consists of a collection of expert rules which are required to meet the control goals. These control rules can be developed from survey results, common sense, general principles and intuitive knowledge. The *IF-THEN* or *IF-AND-THEN* rules are mainly going to be used in designing the controller. The situation for which the rules are projected is given by the IF part. The fuzzy system reaction in this state will be given by the THEN part. The linguistic variables for the moisture and dust control will assume linguistic values and these can be described as shown below.

For the first input variable i.e. moisture with the range 1-5%, see Table 8.

For the second input, dust with the range 0-1% see Table 9.

For the output: the pneumatic valve is a butterfly valve which opens from 0°-90°. When the moisture and dust levels are in the ranges not accepted, the pneumatic valve will close to 0°. For the accepted ranges the valve may open fully open to 90°. The valve can also open to any angle between 0° and 90° depending on the conditions being controlled.

Table 8. Moisture ranges

Crisp Input Range (%)	Fuzzy Variable Name
0-1	Low
1-3	Medium
3-5	High

Table 9. Dust ranges

Crisp Input Range (%)	Fuzzy Variable Name
0-0.02	Acceptable
0.02-0.5	Average
0.5-1	Not acceptable

Figure 22. Fuzzy logic control rule base matrix

Moisture / Dust	Low	Medium	High
Acceptable	Valve is open	Valve is open	Close valve
Average	Valve is open	Valve is open	Close valve
Not acceptable	close valve	Close valve	Close valve

1. IF (moisture is "low") AND (dust is "acceptable") THEN (valve is open)
2. IF (moisture is "medium") AND (dust is "average") THEN (valve is open)
3. IF (moisture is "high") AND (dust is "not acceptable") THEN (close valve)
4. IF (moisture is "low") AND (dust is "average") THEN (valve is open)
5. IF (moisture is "medium") AND (dust is "acceptable") THEN (valve is open)
6. IF (moisture is "high") AND (dust is "acceptable") THEN (close valve)
7. IF (moisture is "low") AND (dust is "not acceptable") THEN (close valve)
8. IF (moisture is "medium") AND (dust is "not acceptable") THEN (close valve)
9. IF (moisture is "high") AND (dust is "average") THEN (close valve)

Nine rules are going to be used by the researcher to control the moisture and dust entering the pneumatic valve of the bottle washer.

Simulation for the Moisture and Dust Control Using Matlab

The Fuzzy Logic Inference system (FIS) editor is shown in figure 23. The FIS offers a convenient access to the other editors for example, the membership editor, rule editor etc, the main importance being on the maximum flexibility to allow for interaction with the FL system. This system is a two inputs and one output system and these systems can perform well as 3D plots are produced which can easily be managed by MATLAB.

The membership function editor helps to display and allows for the editing of all membership functions connected with each variable (input and output) for the entire FIS. In this case the triangular membership function was used.

Figure 23. FIS editor

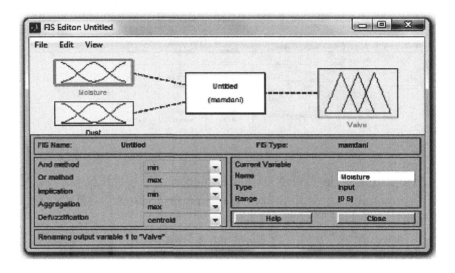

Figure 24. Membership function for moisture and dust

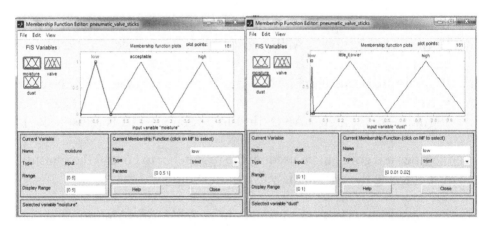

The Rule Editor allows for the editing of a list of rules which define the system behaviour.

Discussion and Analysis of the Effects of Moisture and Dust on Valve

This section will explain the effects of the inputs to the output of the system. These are going to be analysed using the Matlab FLC simulations.

Figure 25. Membership function for Valve and Rule Editor for the FLC

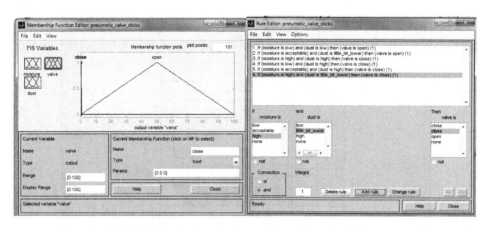

From figure 26 it can be shown that moisture does not have much effect on the valve opening for the short term basis. From 0-5% moisture the valve remains constant. The levels of moisture must not exceed the give ranges as this may result in the failure of the valve. Hence the effects of moisture on valve are minimal as compared to those of dust as shown in Figure 27.

Figure 26. Surface viewer of the valve vs moisture

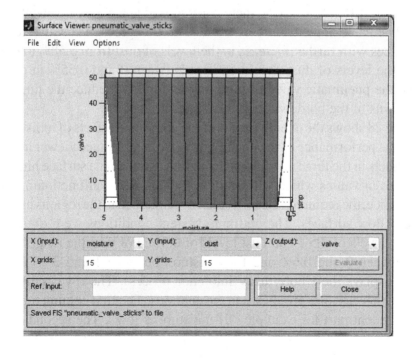

Figure 27. Surface viewer of valve against dust

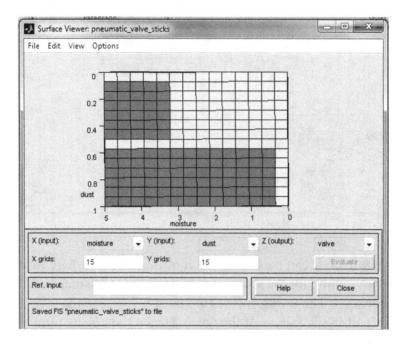

The level of valve opening is maintained at dust levels from 0 to 0.55%. As seen from the surface viewer of valve against dust as the level of dust increases from 0.55% to 0.58%, there is a gradual decrease in the valve opening angle. The valve angle decreases from 45 to 0 degrees. The valve remains at 0 degrees with further increase in the levels of dust from 0.58% to 0.95%. Hence, the levels of dust should be kept between 0 and 0.55% in order to prevent the pneumatic valve from failing and to also reduce the number of breakdowns of the bottle washing machine.

Figure 28 shows the overall results of the combined effects of moisture and dust on the performance of the pneumatic valve. The surface viewer indicates these effects in the three dimensional diagram and the blue surface highlights the ranges of values which will result in the possible malfunctioning of the valve. Hence, the required values should be maintained in the regions indicated by the yellow surfaces. The optimum system conditions were obtained at moisture levels of 2.5% and dust levels of 0.5%. For dust the ranges should vary between 0 and 0.6% and for moisture, between 1 and 3%. The rule viewer displays the whole fuzzy Inference process. The rows indicate the rules and the variables are shown in the columns. The first column represents the antecedent part for example, "If moisture is low". The second column

Figure 28. Surface viewer and rule viewer

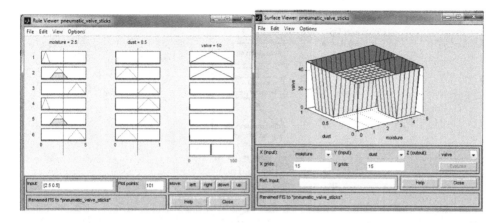

indicates the conditions of dust and the third column displays the response of the valve to the system changing conditions. The resultant aggregate plot is given on the lower right bottom corner of the Rule Viewer under the third column. The red line passing through the aggregate set indicates the defuzzified output value.

The temperatures and pressures inside the bottle washing machine should be kept in the required ranges to avoid unplanned shutdowns and failure of the valve. The temperature ranges in the bottle washer should be between 50 and 85°C. Hence, any range of values < 50 °C and >85 °C may result in failure of the valve and the machine may release unclean bottles. This increases costs as the bottles will be broken when passing through the EBI. For efficient cleaning of the bottles the ranges of temperature should be between 50 and 85 °C. Pressure ranges should also be kept within the required ranges. The linguistic variables for temperature and pressure control will assume membership degree and these can be described as:

For temperature, see Table 10.

For pressure, see Table 11.

The following rules were developed for the pressure and temperature control;

1. IF (temperature is "too low") AND (pressure is "low") THEN (close valve)
2. IF (temperature is "acceptable") AND (pressure is "acceptable") THEN (fully open valve)
3. IF (temperature is "too high") AND (pressure is "high") THEN (close valve)

Figure 29. Temperature and pressure control

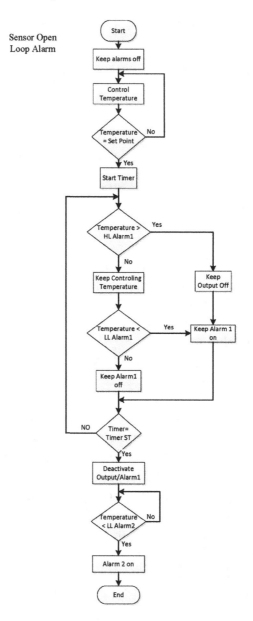

Table 10. Temperature ranges

Crisp Input Range °C	Fuzzy Variable Name
0-50	Too low
50-85	Acceptable
85-100	Too high

Table 11. Pressure ranges

Crisp Input Range (bars)	Fuzzy Variable Name
0-0.5	Low
0.5-3	Acceptable
3-5	High

Table 12. Valve angle

Crisp Range (°)	Fuzzy Variable Name
0	Closed
0-90	Open
90	Fully open

Figure 30. Fuzzy logic control rule base matrix

Temperature / Pressure	Too low	Acceptable	Too high
Low	close valve	Open valve	Close valve
Acceptable	close valve	fully open valve	Close valve
High	close valve	open valve	Close valve

4. IF (temperature is "too low") AND (pressure is "acceptable") THEN (close valve)
5. IF (temperature is "acceptable") AND (pressure is "low") THEN (open valve)
6. IF (temperature is "too high") AND (pressure is "low") THEN (close valve)
7. IF (temperature is "too low") AND (pressure is "high") THEN (close valve)
8. IF (temperature is "acceptable") AND (pressure is "high") THEN (open valve)
9. IF (temperature is "too high") AND (pressure is "acceptable") THEN (close valve)

From the valve angle versus pressure graph the effect of pressure on the valve operation is almost negligible. At 0-5 bars the angle of the valve remains maintained at 45 degrees. The pressure effects are insignificant as compared to the effects of temperature as shown in Figure 35.

From the graph of the pneumatic valve angle of the bottle washer versus the temperature, the temperature affects the valve operation. The valve will be fully open at 90° when the temperature is between 50.1-84.9 °C. As the temperature increases from 0-7 °C the angle of the valve will gradually decrease from 45-0°. The valve will remain closed until the temperature reaches 50 °C. This is because the valve should not operate in this range of

Figure 31. FIS editor

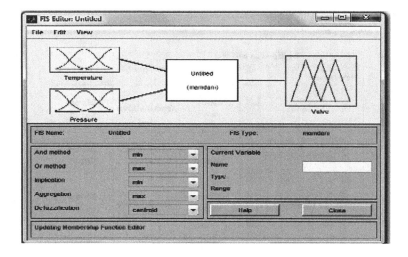

Figure 32. Membership function editor for pressure and temperature

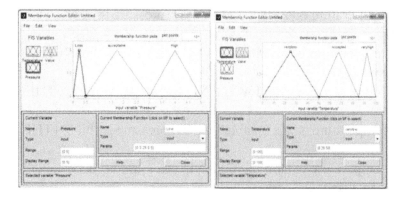

Figure 33. Membership function for valve and rule editor for the FLC

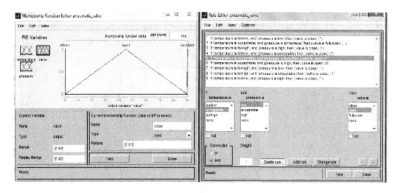

Figure 34. Surface viewer for pressure

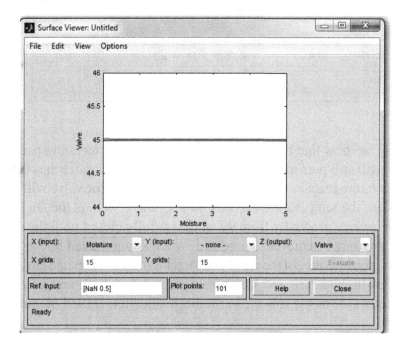

temperature as it could lead to the sticking. The required range for the valve to function optimally is between 50 and 85 °C.

From the Rule Viewer in Figure 35 it can be shown that the valve functions optimally when pressure is 2.5 bars and temperature is 50°C. The surface viewer also shows the regions in which the valve will function efficiently and in this region possible malfunctioning of the valve is reduced. From Figure 36

Figure 35. Surface viewer for temperature

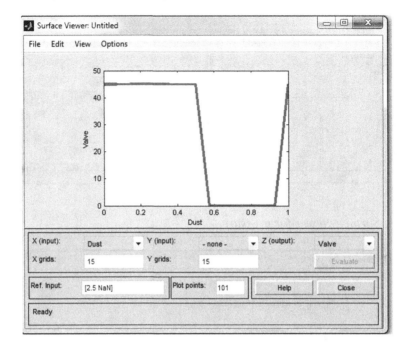

it can also be show that by varying the pressure and temperature parameters the valve will still perform as required. When the pressure reaches 0.51 bars and temperature reaches 50.1°C as shown in Figure 37 the valve will be fully open. When the temperature exceeds 84.9°C and 2.99 bars the angle of the valve will reduce gradually from 90° until it is fully closed until the required range of pressure s achieved. Therefore, by adjusting the red line in the rule viewer the performance of the valve due to changing pressures and temperature can be determined. The red line passing through the aggregate set at the bottom right side of the rule viewer indicates the defuzzified output value.

CONCLUSION

Recommendations for BMC

The FLC model was developed for the valve after the root cause analysis of the problems leading to the sticking and failure. The controller controls the temperature ranges, pressure, dust and moisture leading to the sticking

Figure 36. Rule viewer and surface viewer

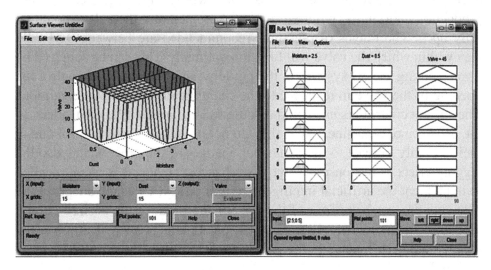

Figure 37. X-Y surface viewer

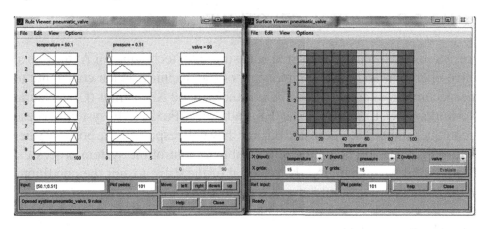

of the valve. From the fuzzy logic rule viewer, it is recommended that the dust should be kept between 0 and 0.55%. Dust can cause scoring and wearing of the valve material and hence this will lead to the failure of the valve. Maintaining the levels of dust in the required ranges can result in the reduced the number of breakdowns of the bottle washing machine. Reducing the number of breakdowns will result in increased production and operating efficiency and thereby reducing losses. It is also recommended that the temperature should be maintained between 50 and 85°C. Temperatures too high or lower than the recommended can lead to the damage of the valve over time hence causing breakdowns which will result in significant loss of production. Within the required ranges of temperature the valve should be fully open. The ranges of pressures recommended are between 0.51-2.99 bars within which the valve will function optimally. This can be achieved by implementing the AI system on the bottle washing machine to control the opening of the valve in response to the changing environmental conditions The current controllers used on the bottle washer are the PID controllers. In the presence of non- linearity's, they do not provide optimal control due to their inability to react to the behaviour changes of the process and also the lagging effect in response to large disturbances. Hence, it is recommended to implement the artificial intelligent system, MRAFC, which adapts to the changing process behaviour of the system and produce the desired output. The Implementation of MRAFC to the pneumatic valve will improve the steady state performance in the presence of non-linearity's or disturbances within the system.

FUTURE RESEARCH DIRECTIONS

Further research should be carried out using other techniques of AI to control valve operation and other components of the machine which contribute to the machine breakdown. These include the Genetic Algorithms (GA), Neural Networks (NN), Expert Systems (ES) and, Neuro Fuzzy (NF) systems. These techniques should each be tested on how they will respond to the bottle washer system changes and how they will be able to control the system processes. The possibility of changing the material of the valve to suit the conditions in the bottle washer should also be considered.

Concluding Ideas

The main objective of this research was to analyse the root causes leading to the sticking and failing of the pneumatic valve of the Krones bottle washing machine and to develop an artificial intelligent system that would help reduce the frequency of sticking and failure of the pneumatic valve and hence, reduce breakdowns of the bottle washing machine. The root cause analysis showed that temperature, pressure, and minor contributions dust and moisture were the main causes leading to the sticking of the valve. The AI system should learn the process dynamics of the valve and be able to adjust the valve angle in response to the conditions. The MRAFC scheme was proposed for the control of the pneumatic valve. A comparison between the MRAFC and MRAC with PID controllers was carried out using simulations and the MRAFC controller proposed showed excellent tracking results as compared to MRAC PID controller. The simulation results showed that transient performance of the system can be improved significantly by the proposed MRAFC scheme.

REFERENCES

Brassington. (2009). *Safe Handling of Tank Containers, International Tank Container Organisation.* ICHCA International Limited.

BSEN415-2:2000. (2015, January 1). *Machinery Standards for Packaging - OYG15*. Retrieved March 12, 2015, from First choice solutions: http://www.machinesafety.co.uk/news/machinery-standards-for-packaging---oyg15

Institute for 21st Century Energy. (2012, December 22). *CO2 Enhanced Oil Recovery*. Retrieved March 12, 2015, from Institute for 21st Century Energy | U.S. Chamber of Commerce: http://www.energyxxi.org/sites/default/files/020174_EI21_EnhancedOilRecovery_final.pdf

Krones. (2012, February 17). Retrieved June 1, 2012, from Krones: http://www.krones.com/en/

Liu, H. J., & Wang, Y. N. (2008). Development of a Computerized Method to Inspect Empty Glass Bottle. *International Symposium on Computer Science and Computational Technology* (pp. 118-123). International Symposium on Computer Science and Computational Technology. doi:10.1109/ISCSCT.2008.340

Mushiri, T. (2012). *A study into the role of fuzzy logic systems in condition based maintenance for control of the pneumatic valve of bottle washer in beverage companies.* Academic Press.

Perry & Green. (1984). *Perry's Chemical Engineers' Handbook.* McGraw Hill.

RockwellAutomation. (2012, February 17). *Solutions action in Columbia machine.* Retrieved March 12, 2015, from Rockwell Automation: literature. rockwellautomation.comidc groups...oem-ap7-en-p.pdf

Shafait, F., Imran, S. M., & Klette-Matzat, S. (2004). Fault detection and localization in empty water bottles through machine vision. IEEE, 30 - 34.

USPolicy. (2013, January 1). *Carbon dioxide enhanced oil recovery: A critical domestic energy, economic, and environmental opportunity*. Retrieved March 12, 2015, from National Enhanced Oil Recovery Initiative: http://www.neori. org/NEORI_Report.pdf

KEY TERMS AND DEFINITIONS

ES: Expert systems.
FLC: Fuzzy logic control.
FMECA: Failure modes effects and criticality analysis.
GA: Genetic algorithms.
MATLAB: Matrix laboratory.
MRAC: Model reference adaptive control.
MRAFC: Model reference adaptive fuzzy control.
NF: Neuro fuzzy.
NN: Neural networks.

Chapter 6

Fuzzy Logic Application in Improving Maintenance in a Mining Company (PMC)

ABSTRACT

At a platinum mining company, the research was based on the root cause analysis technique. The objective was to determine the major causes of failure of the pebble crusher, to estimate between the major crusher failures and provide suitable solutions that included the optimization of the crushing circuit. Major failures were investigated including the breaking of the main shaft, bearing failure, and also entry of tramp iron in the crushing chamber. In solving these problems, analysis of stresses was done using solid works 2015, and condition monitoring techniques were applied using MATLAB 2015 to investigate the development of the crack in the shaft. The results showed that EN 19 has better physical properties than EN 9 and EN 26. EN 19 was recommended for the construction of the main shaft. Crack detection prediction by using MATLAB can be complemented and validated by the use of non-destructive testing.

INTRODUCTION

Platinum is found in nature, in the Earth's crust, in nickel and copper ores with deposits mostly found in South Africa, which produces 80% of the world's platinum. It is rarely, if ever, found in its pure state and has six naturally occurring isotopes (Digby Wells and Associates, 2008).

DOI: 10.4018/978-1-5225-3244-6.ch006

Table 1. PGMs on the periodic table of elements

H																	He
Li	Be											B	C	N	O	F	Ne
Na	Mg											Al	Si	P	S	Cl	Ar
K	Ca	Sc	Ti	V	Cr	Mn	Fe	Co	Ni	Cu	Zn	Ga	Ge	As	Se	Br	Kr
Rb	Sr	Y	Zr	Nb	Mo	Tc	**Ru**	**Rh**	**Pd**	Ag	Cd	In	Sn	Sb	Te	I	Xe
Cs	Ba	*	Hf	Ta	W	Re	**Os**	**Ir**	**Pt**	Au	Hg	Tl	Pb	Bi	Po	At	Rn
Fr	Ra	**	Rf	Db	Sg	Bh	Hs	Mt	Ds	Rg	Cn	Uut	Fl	Uup	Lv	Uus	Uuo
	*		La	Ce	Pr	Nd	Pm	Sm	Eu	Gd	Tb	Dy	Ho	Er	Tm	Yb	Lu
	**		Ac	Th	Pa	U	Np	Pu	Am	Cm	Bk	Cf	Es	Fm	Md	No	Lr

Bold = Platinum group metals

CONE CRUSHER

The major problem currently at the Platinum mine is the cone crusher breakdowns within the production setup. A crusher basically is a machine designed to reduce large rocks into smaller rocks, gravel, or rock dust. These rocks are those ones that have not been crushed by the Semi Auto Genius (SAG) mill. The process is in series SAG and crusher to the ball mill, those rocks above 13mm diameter from the SAG mill are automatically send to the crusher for continuous crushing before they are send to the ball or secondary mill (Telsmith, 2005).

DESIGN OF CONE CRUSHER TO REDUCE BREAKDOWNS WITH FUZZY LOGIC AND CBM AT PMC

A cone crusher is a modified gyratory crusher that has a shorter spindle which is not suspended as in the gyratory crushers but is curved in a universal bearing below the cone (Barry & Napier-Munn, 2006). Cone crushers are used for size reduction purposes. In order to extract the required mineral the ore size has to be reduced for processing. The actual crushing of the ore of the Cone crusher depends with the type of Cone crusher, but mostly size reduction is through compression and attrition. Crushers maybe be classified as stated below:

A crusher can be considered primary, secondary or tertiary crusher depending on the size reduction factor.

Figure 1. Pictorial view of the crusher and details explanation
Telsmith, 2005.

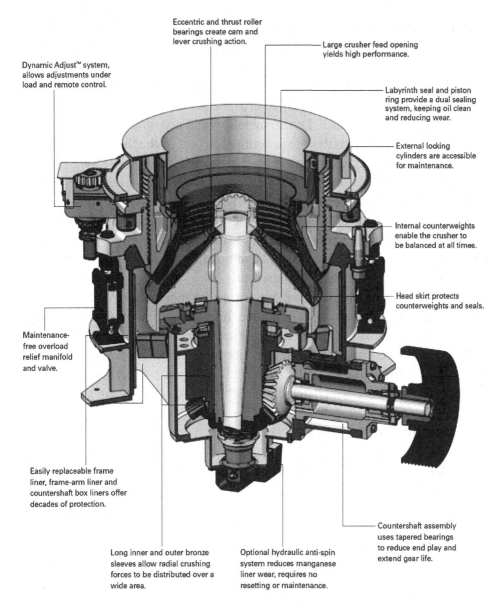

Eccentric and thrust roller bearings create cam and lever crushing action.

Large crusher feed opening yields high performance.

Dynamic Adjust™ system, allows adjustments under load and remote control.

Labyrinth seal and piston ring provide a dual sealing system, keeping oil clean and reducing wear.

External locking cylinders are accessible for maintenance.

Internal counterweights enable the crusher to be balanced at all times.

Head skirt protects counterweights and seals.

Maintenance-free overload relief manifold and valve.

Easily replaceable frame liner, frame-arm liner and countershaft box liners offer decades of protection.

Countershaft assembly uses tapered bearings to reduce end play and extend gear life.

Long inner and outer bronze sleeves allow radial crushing forces to be distributed over a wide area.

Optional hydraulic anti-spin system reduces manganese liner wear, requires no resetting or maintenance.

1. Primary crusher is the one that processes the raw ore straight from the mines. The discharge from the primary crushers is usually channeled to the secondary crushers for further size reduction.
2. Secondary Crusher receive their feed from primary crushers for further size reduction.

Problem Behind PMC

The Frequent Breakdowns of the Installed Cone Crusher at PMC is Costly for Both Maintenance and Production Output

The PMC operates continuously for 24hours. Pebble crusher breakdowns reduce plant throughput. The pebble crusher circuit is used to crush pebbles coming out of the SAG mill. Due to metallurgical accounting facilitated by a metal detector, these pebbles are accounted as part of the throughput. Hence the pebbles are recycled through the pebble crusher circuit. Adding pebbles uncrushed into the SAG mill will reduce the plant throughput because the SAG mill mass also determines the throughput. High SAG mass implies reduced feed rate from the Silos to the SAG mill. Crushing pebbles into fragments of about 14mm by size means that the resident time of these fragments is reduced in the mill. The fragments have no significant impact on the SAG mass and hence the throughput increases.

Increase in the breakdowns will also means an increase in the maintenance costs of the Cone crusher. To realize organization`s profitability, high production at low costs is required. In this regard, pebble crusher optimization is key to improved production and reduced operating costs. Cone Crusher automation helps to foresee breakdowns before they occur and provides online condition monitoring which detects possible causes of failures.

Ore Storage and Reclaim

Currently there are two SILOS in place for storage.

Figure 2. Ore receiving flow

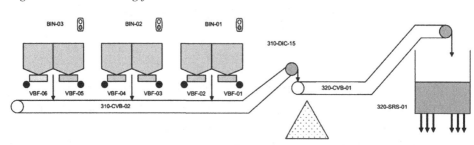

Grinding

The ROM ore in the buffer storage SILOS will be feed in to the SAG mill for comminution using the conveyor belt (330 –MLS -01) via 6 vibrating feeders (FEV-01 -06). For monitoring the SAG mill operations a weightometer is the sensor on the feed conveyor which monitors the feed rate to the mill. The ROM is reclaimed from the buffer SILOS at a nominal rate of 275tph and feed directly into the SAG mill (MLS-01). The SAG mill discharge will be 78%-82% solids. The SAG mill discharge will be graded through a 12mm polyurethane trammel screen to return oversize pebbles back to the SAG mill feed via a conveyor return system (CVB-01-02-03) and a pebble crusher with a closed size setting (CSS) of 13mm. There are a series of three magnets and a cross metal detector before the pebble crusher to remove and detect steel material respectively. The set of magnets serve a function of protecting the pebble crusher from foreign unwanted particles. A low intensity magnet (MDT-01) is along (CVB-01) and it removes tramp steel material mainly steel rods. At the head of the pulley of the (CVB-01) is sited a high intensity magnet (MDT-02) to remove steel material mainly steel balls. The metal detector is also installed just before the pebble crusher so that is detects the non-ferrous materials which might damage the crusher for example Aluminum. A diverter (PNR-01) has been installed to cater for the time when the crusher is down, so that there is no need to stop the operations since the plant is ran 24hours non-stop. The discharge circuit (PNR-01) diverts the pebbles from the circuit to a stockpile. The pebbles are returned to the circuit using loader. The diverter is also used to divert the pebbles from the crusher directly to the crusher discharge conveyor belt in the event that the metal detector has detected presence of a metal .The pebble crusher discharges its product on the conveyor (CVB-03) that discharges onto the SAG mill feed belt. The SAG mill trammel screen undersize will be graded on 6mm deck scalping screen(SCR-01) The 12mm + 6mm material will be returned to the SAG mill feed belt and the -6mm material will be feed to the Ball mill discharge sump via a floatation unit and for further classification and grinding of the cyclone underflow The unit flotation cell (FTF-01) allows recovery of coarse liberated and fast floating sulphides and PGM's to a flotation concentrate as they are produced in the grinding circuit. The Ball mill (MLB-01) operates in closed circuit with a cluster of hydrocyclones (CSC-01- 08). The hydrocyclones classify mill discharge to produce a cyclone overflow at a nominal (P80) sizing of 0.075mm to feed the flotation circuit. Cyclone underflow is gravity fed to the Ball mill feed chute(DIC-08). Cyclone overflow is gravity fed to a linear trash screen (SCL-01) with cloth aperture 1mm for removal of wood chips and other waste material.

Flotation

The main purpose for the floatation stage is to liberate *Platinum Group Metals* and sulphide minerals from the ground ore pulp and produce a concentrate for further processing in a smelter. A linear trash screen (330-SCL-01) has been installed on the cyclone overflow to remove trash before flotation. Linear trash screen undersize from the grinding section at nominal sizing P80 of 0.075mm is fed to a flotation feed stilling box (VES-07) and distributor (DIS-01) which splits the slurry into two equal feed streams. The streams are treated in two identical flotation lines nominated as line A and line B. At this stage flotation tailings is thickened and pumped to a tailings dam and process water recovered from the thickener overflow, clarified in a tailings clarifier and returned to the process water dam.

Figure 3. Primary grinding circuit

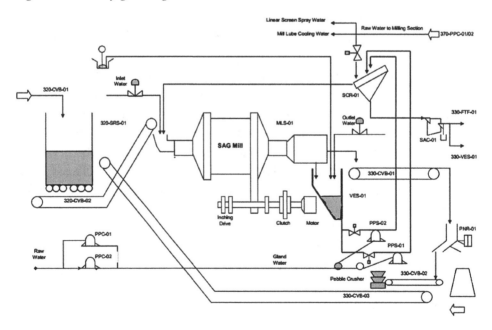

Problems Faced by the Pebble Crusher

A new Osborne Gyrasphere pebble crusher was recently installed in the crushing circuit removing the old one. The old pebble crusher had been installed since 2006. The new pebble crusher has the same closed side setting (13mm) as the old one but the slight deference is that it as a Hydrocone pebble crusher. It maintains its CSS using hydraulics where by when there is tramp iron or any foreign metal inside the crushing the hydraulic system will release pressure inside the system and the main shaft will be lowered allowing the tramp iron to fall or to pass through without damaging the crushing chamber.

Mainshaft Failure

1. The recently installed pebble crusher broke down due to the breaking of the mainshaft. The mainshaft broke separating from the mantle and the concave. From the maintenance team it was discovered that the mainshaft had broken at the neck, and various suggestions lead to the team decide that the cracks had developed gradually until the material gave in. Possible causes of failure of the mainshaft are:

    a. Ingress of water in the crushing system. At PMC wet crushing is practiced, meaning that water is added into the SAG mills so that there will be no dust present in the crushing the addition of water in the circuit will cause the dirt particle to accumulate below the head of the crushing chamber. The presence of the accumulation of the dirt particles will results the dirt particles exerting a compressive force of the mainshaft. The compressive force on the mainshaft will weaken the mainshaft contributing to the formation of cracks.
    b. The presence of the steel balls in the crushing chamber. There are 3 metal detectors in the crushing circuit. One overhead metal detector and 2 magnets which are supposed to remove the steel balls or scraps of metal that might be found in the ore to prevent metal entering the crushing. It is noticed that the metal detectors are not operating at 100% efficiency therefore resulting in the metals finding their way in the crushing chamber. Presence of steel balls will result in damaging of the crushing chamber and also damaging the mainshaft since the pebble crusher will be trying to reduce the size of the metal.

Bearing Failure

On the mainshaft there are two bronze roller bearing which lower the displacement of the mainshaft. The new pebble crusher has recently undergone failure at these two roller contact bearing. The bearing are developed cracks due to the vibrations of the mainshaft. As time goes on the bearing would give in to the subjected impact force and break. The possible causes of the bearing include:

- Large vibrations from the mainshaft result in vibration being induced in the roller bearing. The frequency of vibration would approaching resonance resulting in the cracking of the bearing.
- The clearance distance between the mainshaft and the bearing is too small thus resulting in a higher magnitude of the frequency of the mainshaft being induced in the bearing.

Tramp Iron

The crushing chamber is being damaged by the entry of the foreign metals in the pebble crusher. This results in the crusher trying to reduce the size of the metal in doing so the liner of the crushing chamber will be damaged.

RESULTS AND ANALYSIS OF MAINTENANCE FUZZY LOGIC SIMULATION IN A MINING COMPANY

Introduction

The whole research is based on the concept of Root Cause Analysis (RCA) technique. The whole aim is to reduce the unwanted downtime since the PMC is operated continuously for 24 hours and cut down costly maintenance schedules. The analysis of root cause of failure of the Conical Crusher, consequences that results, identification of the possible solutions to the type of failure and suggesting methods or ways to prevent the same failure from occurring again.

Stress Analysis

It is noted that when a body is in equilibrium then it is subjected to an external load or force internal forces will setup in that material. The intensity if the internal forces setup depends on the load applied /external force applied (Hearn, 1997). There are four types of forces that can act on a body which torsional force, compressive force, torsional force and shear.

Stress = Load/Area

If a bar is subjected to a direct load, and hence a stress, the bar will change in length. If the bar has an original length L and changes in length by an amount +L. This change in the length will result deformation. Strain is thus a measure of the deformation of the material.

Concept Description

Steel balls are entering the pebble crusher's crushing chamber. Since some are not being detected by the low intensity magnets MD-01 and MD -02 and also the metal detector. The pebble crusher will end up trying to downsize the steel ball so that it may passes through the gape. In doing so compressive stresses result since the steel ball are very hard and difficult to crush. The resultant force is induced on the mainshaft thus resulting in the developing of fatigue in the mainshaft. A crack develops as the resultant force will setup internal forces in the steel (En 9) which was used to make the mainshaft.

The Osborne Gyrasphere pebble crusher has the mainshaft made from En 9 steel.

Figure 4. Pebble crusher mantle from Solidworks 2015, Pebble crusher mainshaft from Solidworks, 2015, pebble crushing chamber from Solidworks 2015
Drawn by T. Mushiri.

Material: EN9

Chemical properties are shown in Tables 2 and 3.

Von Mises Calculation

According to the Pebble crusher manual

Table 2. En9 chemical properties

Element	Percentage %
Carbon	0.50-0.60
Silicon	0.10-0.60
Manganese	0.50-0.90
Sulphur	0.050 max
Phosphorus	0.050 max

Table 3. En9 mechanical properties

Element	Percentage %	Element	Percentage %
Carbon	0.50-0.60	Carbon	0.50-0.60
Silicon	0.10-0.60	Silicon	0.10-0.60
Manganese	0.50-0.90	Manganese	0.50-0.90

Table 4. Specification of the selected material EN19 source SOLIDWORKS 2015

Property	Value	Units
Elastic modulus	2.1e+011	N/m2
Poison ratio	0.28	N/A
Shear modulus	7.9e+011	N/m2
Mass density	7700	Kg/m3
Tensile strength	723825600	N/m2
Yield strength	620422000	N/m2
Thermal expansion coefficient	1.3e-005	/K
Thermal conductivity	50	W/m.K
Specific heat	460	J/kg.K
Material damping ratio		N/A

Shaft diameter (d) = 0.46m

Length of shaft (L) = 1.691m

According to the reading on the weightometer

Load = 250 000N

Speed (N) = 1480rpm

Power (P) = 4kW

According to (R.S. KHURMI, 2005)

$$P = \frac{2\pi NT}{60} \tag{1}$$

$$T = \frac{P * 60}{2\pi * N} = 25.08 Nm \tag{2}$$

$$J = \frac{\pi D^4}{32} = 0.004396 m^4 \tag{3}$$

$$Area of shaft = 2\pi r^2 + \pi dl = 2.7761 m^2 \tag{4}$$

According to (Hearn, 1997)
Considering the stress acting in the y direction on the shaft

$$\sigma_y = \frac{Load}{Area} \tag{5}$$

$$\sigma_{y1} = \frac{Load}{Area} = \frac{250000}{2.7761} = 90.05 kPa \tag{6}$$

$$\sigma_{x1} = \frac{Load}{Area} = \frac{250000}{\pi dl} = \frac{250000}{\pi * 0.4624 * 1.691} = 101.772 kPa \tag{7}$$

$$TotalSurfaceAreaOfFrustum = \pi(r + R)s + \pi r^2 + \pi R^2$$
$$= \pi(0.228 + 0.925) * \frac{0.432}{Sin\,45} + \pi(0.228^2 + 0.925^2) \tag{8}$$
$$= 5.064 m^2$$

$$\sigma_{y2} = \frac{250000\,Sin\,45}{SurfaceAreaOfFrustum} = \frac{250000}{5.0643 m^2} = 34.906 kPa \tag{9}$$

$$\sigma_{x2} = \frac{250000\,Cos\,45}{SurfaceAreaOfFrustum} = \frac{250000}{5.0643 m^2} = 34.906 kPa \tag{10}$$

$$TotalAreaThex(\sigma_x) = \sigma_{x1} + \sigma_{x2}$$
$$= 101.772 kPa + 34.966 kPa \tag{11}$$
$$= 136.678 kPa$$

$$TotalAreaThey(\sigma_y) = \sigma_{y1} + \sigma_{y2}$$
$$= 90.05 kPa + 34.906 kPa \tag{12}$$
$$= 124.946 kPa$$

Figure 5. Stresses acting in the crushing chamber
R = 0.925m, r = 0.228m, s = 0.611m

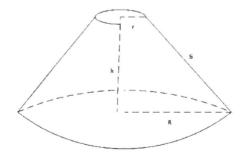

According to (R.S. KHURMI, 2005);

$$\frac{T}{J} = \frac{G\theta}{L} = \frac{t}{r}$$ (13)

where T = torque acting on the shaft, J = polar second moment of inertia, G = Modulus of rigidity, θ = Angular displacement, τ = shear stress, r =radius of shaft

$$\frac{Tr}{J} = t$$ (14)

But T = 25.80Nm

$$t_{xy} = \frac{(25.8 * 0.23)}{0.004396} = 1.35kPa$$ (15)

According to (Hearn, 1997, p. 180);

$$\sigma_1 = \frac{\sigma_x + \sigma_y}{2} + \sqrt{\left(\frac{\sigma_x + \sigma_y}{2}\right)^2 + \tau_{xy}^2}$$ (16)

$$= 132.21kPa$$

$$\sigma_2 = \frac{\sigma_x + \sigma_y}{2} - \sqrt{\left(\frac{\sigma_x + \sigma_y}{2}\right)^2 + \tau_{xy}^2}$$ (17)

$$= -1.823 * 10^6 kPa$$

According to Von Mises

$$\sigma_v = \left[\left(\frac{\sigma_1 - \sigma_2}{2}\right)^2 + \left(\frac{\sigma_3 - \sigma_2}{2}\right)^2 + \left(\frac{\sigma_1 - \sigma_3}{2}\right)^2\right]^{1/2}$$ (18)

$$= 1.338MPa$$

According to (Hearn, 1997, p. 445) using the Söderberg equation

$$\sigma_a = \sigma_N \left[1 - \left(\frac{\sigma_m}{\sigma_y} \right) \right]$$

(19)

where

σ_a = Alternating stress amplitude,

σ_m = Mean stress,

σ_N = Fatigue strength for N cycles,

σ_y = Yield strength of material.

From material specifications

EN19 σ_y = 620 MPA

EN26 σ_y = 720 MPA

$$\sigma_m = \frac{1}{2} \left[132.21 * 10^3 + -1.823 * 10^6 \right]$$
$$= -845.395 kPa$$

(20)

$$\sigma_a = \frac{1}{2} \left[\sigma_{max} - \sigma_{min} \right]$$
$$= \frac{1}{2} \left[132.21 * 10^3 - -1.823 * 10^6 \right]$$
$$= 977.605 kPa$$

(21)

σ_N = fatigue strength for N cycles for EN19

$$\sigma_a = \sigma_N \left[1 - \left(\frac{\sigma_m}{\sigma_y} \right) \right]$$
$$977.605 * 10^3 = \sigma_N \left[1 - \left(\frac{-645.395 * 10^3}{620 * 10^6} \right) \right]$$
$$\sigma_N = 976.273 kPa$$

(22)

σ_N = Fatigue strength for N cycles for EN26

$$\sigma_a = \sigma_N \left[1 - \left(\frac{\sigma_m}{\sigma_y} \right) \right]$$

$$977.605 * 10^3 = \sigma_N \left[1 - \left(\frac{-645.395 * 10^3}{720 * 10^6} \right) \right] \tag{23}$$

$$\sigma_N = 976.458 kPa$$

For the design to be safe the $\sigma_v \leq \sigma_y$

σ_y is the yield strength of the materials

$$\sigma_y = 620 MPa \ / \ 720 MPa$$

AUTOMATION

For the $\frac{1}{2}$ '' sieve size

Weight meter feed conveyor = 250tph

Figure 6. Loading conditions, forces acting on the crushing chamber

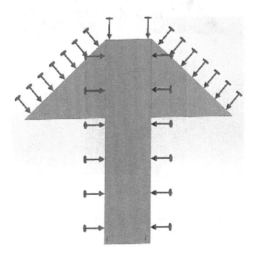

Figure 7. Von Mises results

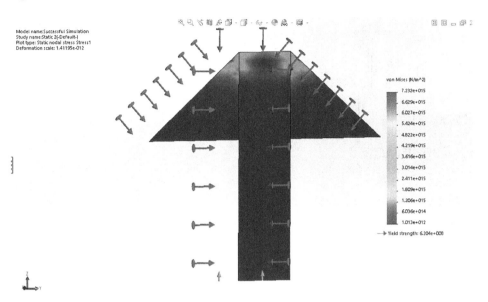

*For a more accurate representation see the electronic version.

Figure 8. Strain results

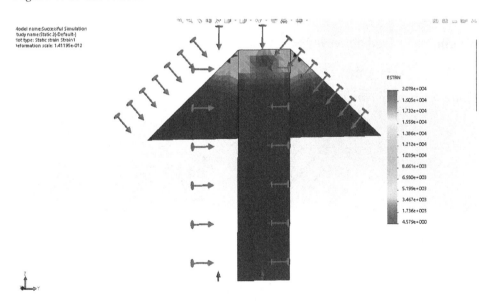

*For a more accurate representation see the electronic version.

Figure 9. Stress results for EN26

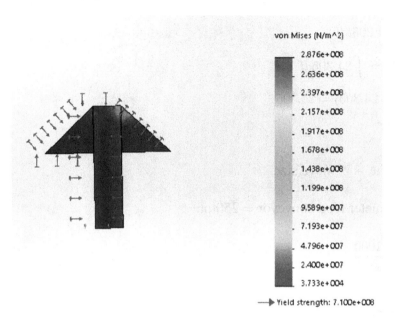

*For a more accurate representation see the electronic version.

Table 5. Mass distribution

Sieve Size	CSS	Flow Rate/ Kgs^{-1}
$\dfrac{1}{2}$ ''	12.7mm	35%
$\dfrac{3}{8}$ ''	9.525mm	26%
$\dfrac{3}{4}$ ''	19.05mm	52.5%

$$= \frac{250 * 1000}{3600} = 69.44 \text{kgs}^{-1}$$

Considering the 35% passing $= \dfrac{35}{100} * 69.44$

$= 24.305 \text{kgs}^{-1}$

$$\frac{dM}{dt} = 24.305t$$

$$dM = 24.205tdt$$

$$\int_{24.305}^{69.44} dM = \int_{0}^{t} 24.305dt \qquad (24)$$

$$(69.44 - 24.305) = 24.305t^2$$

$$t = 1.93s$$

For the $\frac{3}{8}$'' sieve size

Weight meter feed conveyor = 250tph

$$= \frac{250 * 1000}{3600}$$

$$= 69.44\text{kgs}^{-1}$$

Considering the 26% passing $= \frac{26}{100} * 69.44$

$$= 18.055\text{kgs}^{-1}$$

$$\frac{dM}{dt} = 18.055t$$

$$dM = 18.055tdt$$

$$\int_{18.055}^{69.44} dM = \int_{0}^{t} 18.055dt \qquad (25)$$

$$(69.44 - 18.055) = 18.055t^2$$

$$t = 2.385s$$

For the $\frac{3}{4}$'' sieve size

Weight meter feed conveyor = 250tph

$$= \frac{250 * 1000}{3600}$$

$= 69.44 \text{kgs}^{-1}$

Considering the 52.5% passing $= \dfrac{52.5}{100} * 69.44$

$= 52.083 \text{kgs}^{-1}$

$\dfrac{dM}{dt} = 52.083t$

$dM = 52.083t dt$

$$\int_{52.083}^{69.44} dM = \int_{0}^{t} 52.083 dt \tag{26}$$

$(69.44 - 52.083) = 52.083t^2$

$t = 0.816s$

It is noted that the size distribution has an exponential decrease with time. For analysis using Simulink, the $\dfrac{1}{2}$ '' sieve was used.

Fuzzy Logic Controller

The specification of the pebble crusher are given below:

Speed= 1480rpm

Power = 4kW

Sensor = velocity transducer

According to the academic research done by (Mr.S.P.Bhide, 2015). It was observed that there is 33% increase in amplitude of acceleration of cracked shaft. The velocity transducer is attached on any position on the shaft since all the point have a uniform velocity.

Edraw Max Results

The causes of failure of the pebble crusher are represented using the 6 sigma.

Table 6. Monitoring parameters

Crack Position	Crack Depth/cm	Condition	Speed Range/rpm
No crack	0	Normal	1470-1480
Initial stage	5-10	Critical	1470-1970
Propagation	10-40	Warning	1960-2460

Figure 10. Input variable and output variables

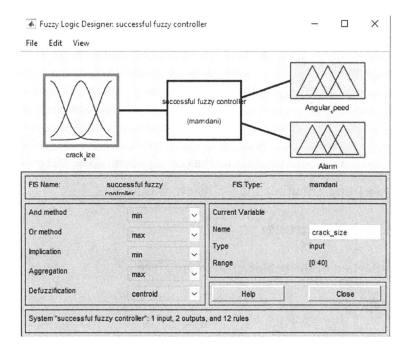

Metal Detector

There are a series of three magnets and a cross metal detector before the pebble crusher to remove and detect steel material respectively. The set of magnets serve a function of protecting the pebble crusher from foreign unwanted particles. A low intensity magnet (MDT-01) is along (CVB-01) and it removes tramp steel material mainly steel rods. At the head of the pulley of the (CVB-01) is sited a high intensity magnet (MDT-02) to remove steel material mainly steel balls. The metal detector is also installed just before the pebble crusher so that is detects the non-ferrous materials which might

Figure 11. Crack propagation membership functions

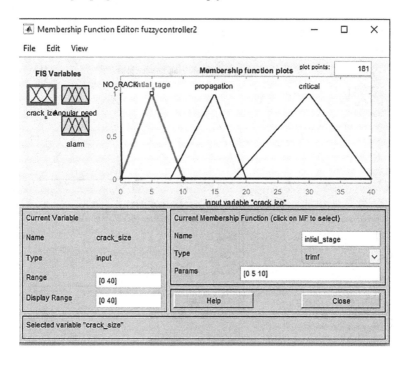

Figure 12. Angular speed membership functions

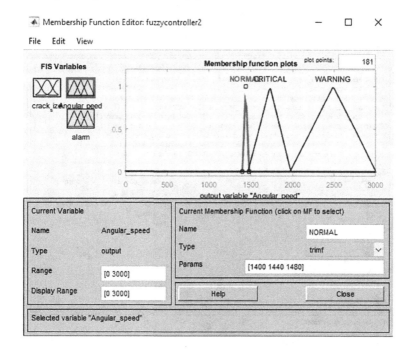

Figure 13. Alarm system membership function

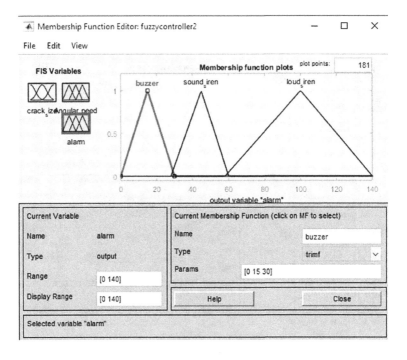

Figure 14. Rules that control the fuzzy controller

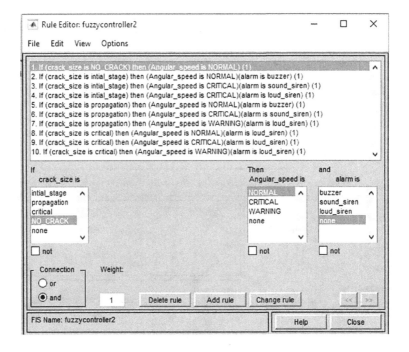

Figure 15. Angular speed against crack propagation

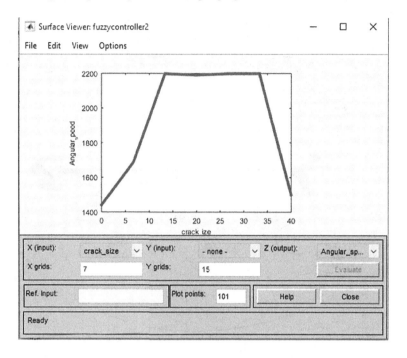

Figure 16. Alarm against crack propagation

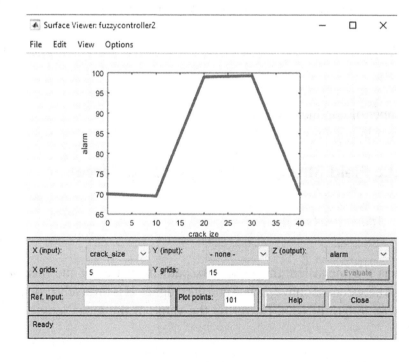

Figure 17. Causes of failure of the pebble crusher

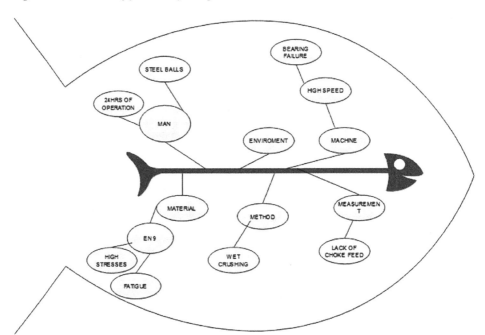

damage the crusher for example Aluminum. Magnetomotive force (mmf) is, the cause of the existence of a magnetic flux in a magnetic circuit:

$$F_{mmf} = N1(amperes),$$ (27)

N = Number of conductors and
I = Current in amperes.

Magnetic Field Strength (or Magnetizing Force)

$$H = \frac{NI}{l}(Ampereperturn)$$ (28)

where l is the mean length of the flux path in metres or the distance of separation between the conveyor belt and the magnet. H is inversely proportional to the mean length of flux. Lowering the value of the mean length of flux increases the magnetic field strength (H), thus more ferrous particles can be detected by the magnet.

Currently the circuit arrangement is like shown below, L = 300mm, Size of particles on the CVB01= 60mm, Safety factor = 2. Equation 28 holds for l = 300.

Considering that the feed from the SAG mill has diameter of 60mm

$$H_f = \frac{NI}{l * safetyfactor}$$

$$= \frac{NI}{60 * 2} \qquad ; = 66.63\% \qquad (29)$$

$$Efficiency = \left[(H_f - H_i) \Big/ (H_i * 100\%) \right]$$

INTERPRETATION OF RESULTS ON FINITE ELEMENT OF THE CRUSHER

Von Mises

In this research two approaches were conducted to find out will the material selected be able to withstand the applied load. The alloy selected for the Stress analysis using Solidworks has the best characteristics in terms of yield stress.

Figure 18. Magnetic circuit arrangement

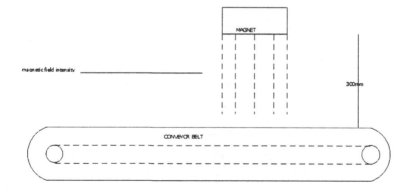

Regions represented in the color code orange and red (6.027e+015-7.232e+0.15) MPa are the region of the high stresses.

- These stresses are found to be acting at the neck of the shaft in the crushing chamber.
- And in this region, fatigue of a the material is likely to develop and crack propagation will yield if the yield strength is exceeded
- The yield stress of the material during the loading conditions of the pebble crusher

$$= 6.204e+008MPa$$

If this yield stress is exceed the material will start to undergo fatigue which will lead to crack development

The regions represented in the color code yellow and green (5.424e+015-3.014e+015) MPa experience moderate stresses.

Figure 19. Extract from Solidworks material EN19

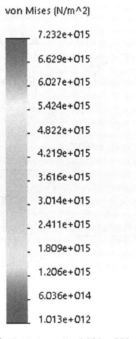

von Mises (N/m^2)

7.232e+015

6.629e+015

6.027e+015

5.424e+015

4.822e+015

4.219e+015

3.616e+015

3.014e+015

2.411e+015

1.809e+015

1.206e+015

6.036e+014

1.013e+012

▶ Yield strength: 6.204e+008

For a more accurate representation see the electronic version.

- These stresses are found to be developing on the upper region of the crushing chamber

The regions represented in the color code light blue and dark blue (2.411e+015- 1.013e+012) Mpa are areas of the crushing chamber that experience low stresses.

- The areas of in the lower region of the crushing chamber experience low stresses during the loading conditions.
- Failures are unlikely to occur in this region.

Fuzzy Logic Controller

There is an increase in velocity by 33% when the crack is propagating in a shaft. Since the motor being used to power the pebble crusher is variable speed drive. The normal operating speed is 1440 rpm. According to the graph when the crack propagates the speed is also increasing which will make it easier for the controller at the SCADA to recognize the development of crack and also when the crack propagates above 50% of the shaft diameter the shaft will break off and stop. The alarm system goes hand in hand with change in speed of the shaft. Change in speed of the shaft is recognized by accelerometer and the change in strain of the will produce an electric charge that will trigger the alarm system.

Metal Detector

Magnetic field strength is affected by the medium in which the magnetic field is travelling in. different mediums have different permeability of free space. At this case the medium is air thus:

$$\mu_0 = 4\pi * 10^{-7} H \, / \, m \tag{30}$$

The magnetic field strength decreases with distance from the source of magnetism. Decreasing the distance increased the efficiency of magnet by 66.63%.

RECOMMENDATIONS OF THE PMC

In this research it can be seen that the main failures of the pebble crusher are oriented about the mainshaft. In the research presented here there is much development need to meet the required performance of the system. Further research has to be done in the following lines:

- **Crack Detection in the Shaft:** The method used in this research is not the best as the change of speed of the shaft can be caused by other variables. Therefore to detect the cracks in the shaft non-destructive tests (NDT) have to be done during the maintenance schedules which are Radiography (RT), Ultrasonics (UT) and Liquid Penetrant (PT)
- **Von Mises:** In this research two materials were analyzed EN 19 and EN26. For machination of the shaft EN 19 is highly recommended since in terms of cost it is cheaper than EN26 although they serve the same function. The difference in the yield strength of the materials is very small, under the loading conditions they both are able to endure the loading conditions.
- **Magnetic Intensity:** In this research the distance of separation between the magnet and the conveyor belt was reduced so that magnetic intensity is increased. This was done so that tramp iron which was passing through can be detected but however if more researches are done the speed of the conveyor belt that is channeled to the pebble crusher can be reduced so that the efficiency of the magnet can be increased
- **Cost:** It is wiser to change the mainshaft than to change the roller bearing because the failure of the bearing is caused by the displacements of the shaft. When the problem of the shaft is solved the roller bearing will not be changed. Since the bearing failure occurred twice in a space of a year, the prices of bearing cost is lower but in the long run it is wiser to fabricate a new shaft made from EN 19 to avoid the bearing failure costs.

CONCLUSION

SMC Concentrator has increased its production throughput after the installation of the pebble crushing circuit in 2008. There is potential to increase the throughput by 5% if breakdowns on the crusher are reduced .Therefore these is great need to optimize the pebble crusher since it is found in the critical path of the whole production path.

REFERENCES

Barry, W. A., & Napier-Munn, T. J. (2006). *Will Mineral Processing Technology* (7th ed.). Oxford, UK: Butterworth-Heinemann.

Bhide, P. (2015). Effect of crack depth of Rotating stepped Shaft on Dynamic. *International Research Journal of Engineering and Technology, 2*(6), 1-5.

Digby Wells and Associates. (2008). *GDACE Mining and Environment al Impact Guide, Mining and Environmental Impact guide.* Johannesburg, South Africa: Gauteng Department of Agriculture, Environment and Conservation.

Hearn, E. J. (1997). An Introduction to the Mechanics of Elastic and Plastic deformation of Solids and Structural Materials. In *Mechanics of Materials 2* (p. 445). Oxford, UK: Butterworth-Heinemann.

Khurmi, J. G. (2005). Torsional and Bending Stresses in Machine Parts. In A textbook of machine design (pp. 120-180). Eurasia Publishing House (Pvt.) Ltd.

Telsmith. (2005). *SBS cone crushers.* SBS Cone Crushers.

KEY TERMS AND DEFINITIONS

PGM: Platinum group metals.
SAG: Semi-auto genius mill.

Chapter 7

Fuzzy Logic Application and Condition Monitoring of Critical Equipment in a Hydro Power Generation Company

ABSTRACT

In Hydropower generation company, the focus was on making use of the reliability-centred maintenance (RCM) principles in relation to expert systems in order to optimize maintenance of power generation assets at HPGC station. The researchers realised that in order for CBM to come out clearly, it is critical to do the RCM first. Naturally, the ageing equipment demands a paradigm shift in maintenance strategies in order to guarantee continuity of supply and meet the ever-growing stakeholder requirements. Increase in customer demand has worsened the situation. There is need to improve the overall equipment effectiveness (OEE) from the current 60% to the world class 85%, using the same old equipment in order to retain customer satisfaction.

INTRODUCTION

Power generation business depends heavily on machinery of high monetary value. Very little study has been done on Reliability Centred Maintenance in the context of developing electricity industry. As a result of this research deficiency, the power generation industry has suffered due to the shortcomings

DOI: 10.4018/978-1-5225-3244-6.ch007

of the traditional maintenance practices of fixed overhauls and breakdown maintenance. This book is going to make use of the Reliability Centred Maintenance (RCM) principles in relation to Expert Systems in order to optimize maintenance of power generation assets at HPGC station. The researchers realised that in order for CBM to come out clearly, it is critical to do the RCM first. Naturally, the ageing equipment demands a paradigm shift in maintenance strategies in order to guarantee continuity of supply and meet the ever growing stakeholder requirements. Increase in customer demand has worsened the situation. There is need to improve the Overall Equipment Effectiveness (OEE) from the current 60% to the world class 85%, using the same old equipment in order to retain customer satisfaction.

Of late, numerous breakdowns together with extended Mean Down Times (MDT) have negatively impacted the company's competitive performance objectives (CPOs) and the Zimbabwean economy at large. The failure rate stands at about 30% against a world class of 10% while mean down time goes to 2 months a year per machine which can be reduced by half with improved maintenance systems. As such, system failure has caused massive loss of revenue amounting to at least $1m a year as well as catastrophic loss of capital equipment worth over $100 000 a year, posing danger to safety, health and the environment. Constant machine overhauls have not improved equipment reliability significantly. The massive exodus of Engineers has added onto the problems leaving the organisation exposed due to lack of experts and hence the need to adopt a streamlined Artificial Intelligence based reliability maintenance system. The aim of this study is to develop a Reliability Centred Maintenance system based on Expert Systems. This is so as to adopt a maintenance approach which would optimise power generation processes at HDPC. The project objectives are:

1. To apply the concept of Maintenance Free Operating Period (MFOP) reliability metric to determine maintenance intervals as opposed to Mean Time Between Failure (MTBF).
2. To determine and analyse the failure mode and effects of equipment and component failure.
3. To design an online condition monitoring scheme for critical equipment.
4. To design a predictive maintenance system as a component of reliability based maintenance.

Figure 1. Flow diagram of a power plant

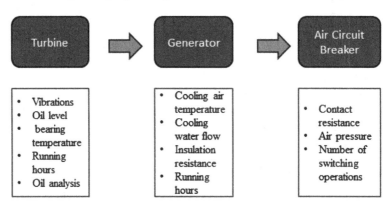

The book focuses on the maintenance of power generation assets at HDPC which includes the hydro-turbines, generators and switchgear. One out of six units only is going to be examined and the concepts can be easily duplicated to the other units. Unit 1 is going to be used for the purpose. The current maintenance practices shall be studied and the Expert Reliability Centred Maintenance system shall be applied as an improvement to the current maintenance systems. The project excludes the transmission and distribution network.

CONTEXT DIAGRAM

Reliability-Centred Maintenance (RCM) integrates Preventive Maintenance (PM), Predictive Testing and Inspection (PT&I), Repair, and Proactive maintenance (NASA, 2008). If these components are properly integrated, machine data is gathered online using sensors, off line maintenance tests data collected and machine operating data collated, an optimized RCM is realised. As shown in Figure 2, the Reliability Centred Maintenance system will be based on six modules. The online Condition monitoring system will have sensors as the primary automatic data collectors. The data is relayed to an S7-1200 and OPC server for processing. Information, advice and explanations on maintenance are relayed to the Maintenance Planner via the HMI and the computer. Simpler maintenance like lubrication shall be done by the actuators. The Ms Access database captures all the information regarding the other modules for further processing by using the required relationships

Figure 2. Context diagram

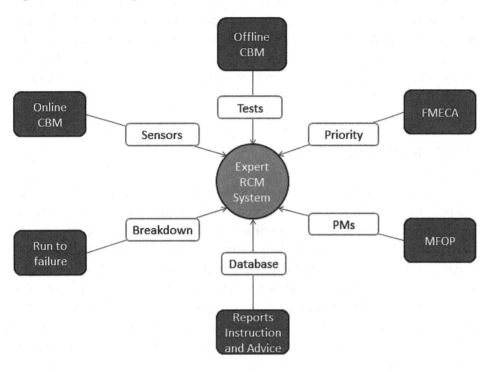

between entities. The diagram seeks to illustrate how the optimisation of different maintenance strategies is to be achieved.

The business of any power station relies on equipment reliability. This is because power stations are asset intensive entities and as a result requires more attention on the equipment care side than in the past. The need to optimize Enterprise Asset Management processes has been on the increase as the generation assets in the field continue to become more unreliable due to age. Replacement of such assets would be capital intensive and won't be an easy exercise. The only prudent alternative is to make do with what is available while tapping the best out of the exploitation. Online Condition Monitoring systems have been developed but what is required is a more detailed approach where a system which optimises the full package of expert maintenance systems is developed rather than concentrating on online monitoring or Computerised Maintenance Management System (CMMS). There have been major breakdowns at HDPC as a result of inadequate maintenance of equipment compounded by the age of the plant. This has resulted in major financial losses amounting to over $1m a year to the business unit. In 2012 alone, $700000 was used to replace failed transformers and hence the need of a more comprehensive and objective approach to maintenance of the physical assets. Unplanned maintenance have driven the availability of the power station to as low as 60%. This may be roughly estimated at over 30% loss in revenue. Further analysis to the challenges reflect on the lack of the appropriate experts to deal with maintenance problems and hence the need of ERCM. ERCM is a component of fuzzy logic in question to this research.

RESEARCH OBJECTIVES

The objectives of this research methodology are to outline the various equipment parameters to be monitored. Maintenance strategies applied on the equipment includes online condition based maintenance combined with the offline predictive testing and inspection trending, equipment replacement, Maintenance Free Operating Periods concepts, fixed time maintenance and corrective maintenance. The equipment data shall be analysed by the controller and the computer. Maintenance decisions shall be done by the logic (Expert RCM system) in the controller and computer. The main outputs of the expert system are maintenance instructions, advice, explanations, maintenance information and maintenance actions.

A comparison on the benefits of the new Expert RCM system with the current Overhaul strategy at Kariba South Power Station shall be done. Parameters like Availability, Overall Equipment Effectiveness, MFOP, costs and revenues shall be used to compare the two systems to assess the gains to be brought by the proposed Expert system RCM.

Power stations are a critical part of the economy of any country. Continuous improvement (KAIZEN) in the field of maintenance is required to ensure optimum plant operations so as to improve availability beyond world class standards. Power plant maintenance can be automated beyond online condition monitoring or computerised maintenance management systems by using current developments in other industries like the MFOP application in the airline industry as well as the rapid growth of computer power and capabilities.

THE HYDRO-ELECTRICITY GENERATION PROCESS

Figure 3 shows a typical hydro electricity generation process. This process is the same arrangement as at HPDC.

Figure 3. Hydropower generation process

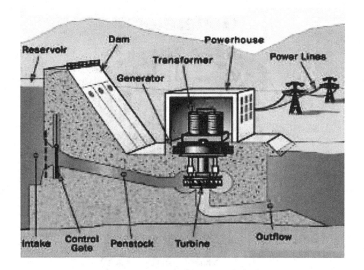

Hydro-Generator Components

The Figure 4 shows the major components of a Hydro-turbine system. The hydro-generator is the most expensive part of the generation system in terms of value. This sub system forms the heart of the generation process and hence key and accurate maintenance decisions are required to ensure optimum operations. Too often fixed time maintenance is as bad as inadequate maintenance. Failure of the machine results in long outages and the diagnosis and repair process are expensive and the expertise is normally outsourced. Intelligent Systems can be easily applied to optimize the maintenance processes.

The components and their relationships are as shown in the graph in Figure 5.

The system is composed of the components as shown. There are two redundant parallel Exciters for the excitation of the Unit (generator). One exciter is duty while the other is on 'hot' standby. The standby exciter has the capability of taking over the full function of the duty in case of failure. The machine can be reconfigured by means of load reduction if parallel components not necessarily in a redundant fashion such as coolers and brushes fail. These decisions are done by the engineer and can be designed into an ERCM system.

Figure 4. The turbine-generator diagram

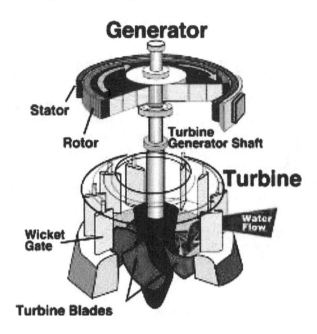

UNIT 1 PARETO ANALYSIS

Failure data for the unit under study (Unit 1) was extracted from SAP and the summary of the maintenance costs for the unit from April 2000 to February 2014 is as shown in Appendix 1. It can be seen from the graph in Figure 6 that the thrust bearing alone contributed 25% maintenance costs and loss of revenue. The thrust bearing contributed over 60% of unit stoppage in the period under review.

Thrust Bearing Pareto Analysis and Failure Modes

The relationship of the thrust bearing to the rest of the system is as shown in Figure 6. The thrust bearing does not have redundancy and failure would directly affect the functionality if the system marked in purple. Between year 2000 to 2014, failure due to bearing vibrations and overheat contributed over 60% of the bearing maintenance costs while they contributed to over 25% of the unit stoppages caused by thrust bearing failure, the distribution from this analysis has led to the idea of concentrating on the bearing temperature

Figure 5. System components

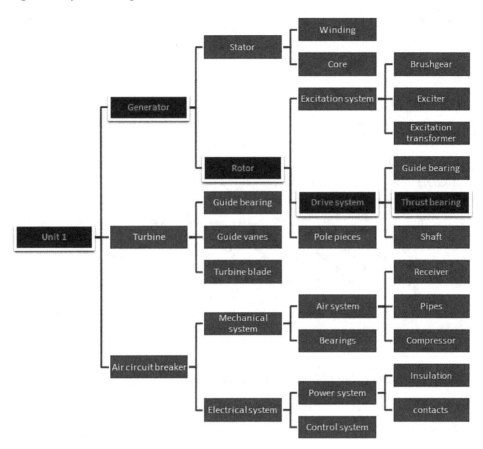

and vibration monitoring as well as the oil level. The other modes could be grouped as depicted in the fishbone diagram. The data used to construct the Pareto chart in Figure 7 is detailed in Appendix 1 and was queried from SAP.

The failure mechanisms of the other components were drawn from the failure data available. Stator winding failure modes are as shown graphically in Figure 8. Similar analysis was done to the rest of the components and the exercise helped in producing the FMECA by identifying the failure modes, causes, effects and remedial action.

HPDC uses the three maintenance strategies indicated in Figure 9. However, the strategies need coordination and integration for optimum maintenance decision making. An Engineer generally asks himself the following questions when making maintenance decisions:

Figure 6. Unit 1 Pareto Chart

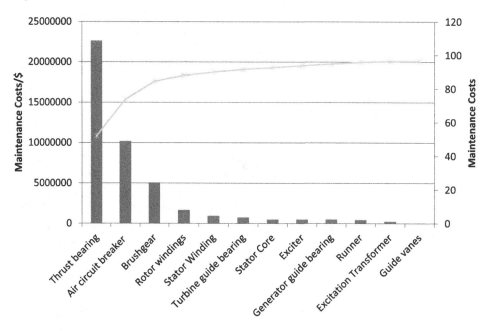

Figure 7. Thrust bearing failures Pareto chart

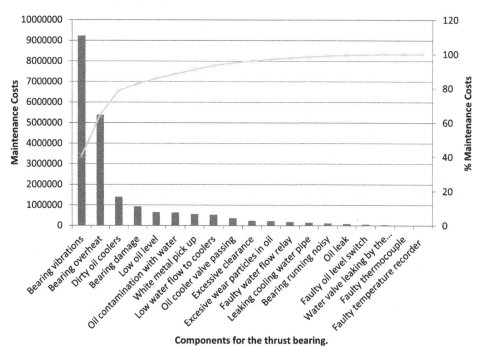

Figure 8. Stator winding failure statistics

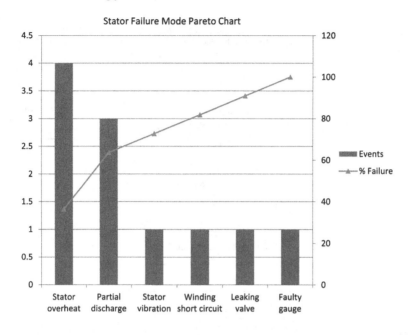

1. What maintenance needs to be done? Maintenance scope
2. Why should it be maintained? Justification of the action
3. When should maintenance be done? Interval of maintenance
4. Which equipment is to be maintained? The equipment to be maintained
5. How should the maintenance be done? Procedure

EXPERT RELIABILITY CENTRED MAINTENANCE SYSTEM AND FUZZY SYSTEM DESIGN FOR A HYDRO POWER GENERATION COMPANY

Introduction

This section details the design of the Expert RCM System using the S7 1200 and Ms Access. The chapter will illustrate the formulation of the algorithms for the thrust bearing maintenance. The incorporation of fuzzy reasoning in the Expert System helps to handle the "vagueness, ambiguous, qualitative, incomplete or imprecise information" associated with plant maintenance. RCM use linguistic statements such as "oil level is very low" or "priority is

Figure 9. The Maintenance decision making process

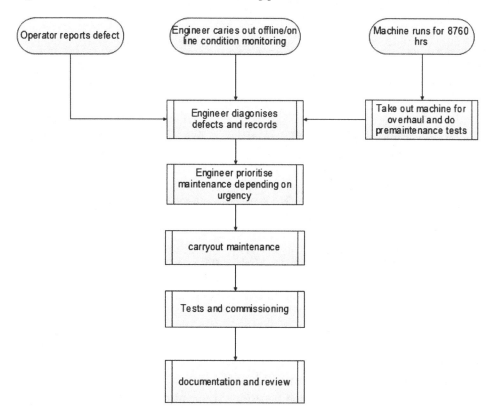

urgent" to commit resources to carry out maintenance of the plant. Figure 10 shows the flow chart used to develop the code on the S7 controller and the Ms Access database.

Rule Formulation Process

The results of the FMECA process and Pareto analysis were used to come up with the rules. Fishbone diagram in Figure was also used to define the causes of the various system defects. Figure 11 shows the fishbone diagram which was constructed using historical equipment maintenance data and expert judgement.

Figure 10. The Expert System flow chart

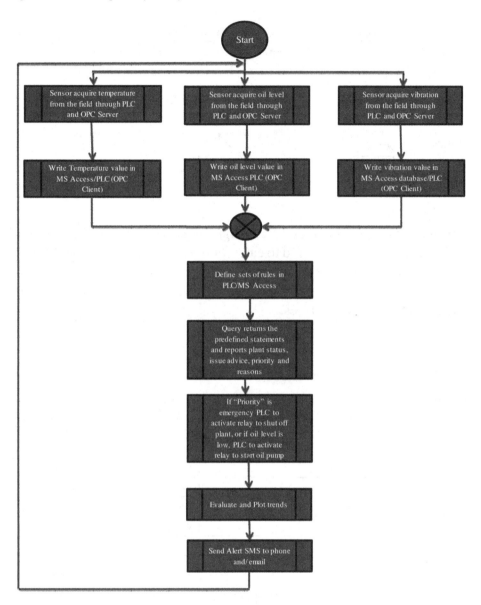

Fuzzy Sets

From the Pareto analysis, the researcher considered monitoring three thrust bearing variables affected by the bulk of the system failure modes; these are temperature, vibrations and oil level. The instances of the variables were

Figure 11. Fishbone diagram

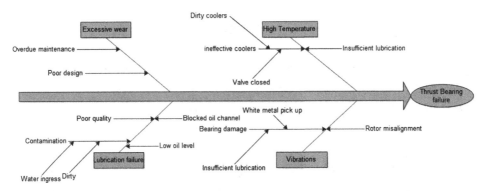

defined as *'normal'*, *'high'* and *'very high'* as shown in the Table 1. A total of 27 rules (3^3) were considered to cater for the combinations of the different states of the sets to make maintenance decisions. It can be noted that the system can be easily expanded to accommodate more sets and a proportional increase in the number of rules.

Table 1. Crisp and fuzzy input sets

	Variable	Range	Fuzzy Set.
Temperature	Temp1	T<80	Normal
	Temp2	80<=T<100	High
	Temp3	T>=100	Very high
Vibrations	Vib1	V<4	Normal
	Vib2	4<=V<5	High
	Vib3	V>=5	Very high
Oil level	Level1	L<1	Very Low
	Level2	1<=L<2	Low
	Level3	L>=3	Normal

where:
- Temp1.........a set of normal temperature levels
- Temp2.........a set of High temperature levels
- Temp3.........a set of Very high temperature levels
- Vib1.............a set of Normal vibration levels
- Vib2.............a set of High vibration levels
- Vib3.............a set of Very high vibration levels
- Level1..........a set of very low oil levels
- Level2..........a set of low oil levels
- Level3..........a set of Normal oil levels

System Crisp Inputs

A total combination of 27 sets (3^3) of temperature, vibration and oil level can be interpreted to represent the various machine states as well as the corresponding actions which need to be taken. Table 2 shows the crisp input combinations from which the rules ware formulated.

Initial Rules

The complete set of the initial rules are listed in Appendix below.

1. **IF** Temperature is Normal **AND** Vibration is Normal **AND** Oil level is Normal **THEN** Machine is running normally **(SEE TRENDS) AND** Priority is Normal **GOTO** Start **ELSE** Next rule
2. **IF** Temperature is Normal **AND** Vibration is Normal **AND** Oil level is Low **THEN** Bearing oil needs top-up, oil pump running (PLC) **AND** Priority is Deferred **GOTO** Start **ELSE** Next rule
3. **IF** Temperature is Normal **AND** Vibration is Very High **AND** Oil level is Very Low **THEN** Check oil leakage and bearing surface for white metal pick up **AND** Priority is Urgent **GOTO** Start **ELSE** Next rule

Table 2. Crisp Inputs combinations

Fuzzy Sets		Oil Level
Temperature	**Vibration**	
Temp1	Vib1	Level3
Temp1	Vib1	Level2
Temp1	Vib1	Level1
Temp1	Vib2	Level3
Temp1	Vib3	Level3
----------------	-------------	-------------
Temp1	Vib3	Level2
Temp1	Vib3	Level1
Temp3	Vib3	Level3
Temp3	Vib3	Level2
Temp3	Vib1	Level1
Temp3	Vib1	Level2

....

....

27. **IF** Temperature is Very High **AND** Vibration is Very High **AND** Oil level is Very Low **THEN** Mission failure imminent, unit shutting down(PLC) **AND** Priority is Emergency **GOTO** Start **ELSE** Next rule

The PLC Programme

1. **Machine Shut Down Under Emergency Conditions:** The PLC was programmed using the ladder shown in Figure 12 shows the logic which shuts off the machine in emergency cases. In this case the Thrust bearing temperature is *'High'*, Thrust bearing vibration is *"very High"* and oil level is *"very low"* and the machine is shut down by the PLC and at the same time sounding the Emergency buzzer and light to alert the operator. This condition is recorded in Ms Access and advice that the machine has been shut down and the cause of the shutdown is issued to the user.

Figure 12. Machine shutdown logic

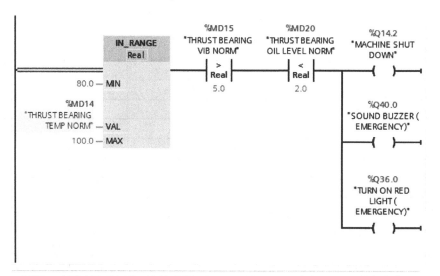

2. **Oil Top Up:** In some cases, the Expert system decides to top up thrust bearing oil when it is low and all the other conditions are normal, the PLC runs the oil pump automatically. The code is as indicated in the rung in Figure 13 the associated alarms are required for the benefit of the operator manning the equipment. The condition is analysed by the expert system and the appropriate advice and actions are issued.

3. **OPC Server:** Three tables were created in the OPC Client (Ms Access) for recording temperature, vibrations and oil level as measured by the PLC through the sensors. Each value of the variables from the PLC is written in a specific field of a table in Ms Access with a time stamp. The time stamp is used as the primary key for retrieval and further processing. Figure 14 shows the temperature values written in Ms Access.

The three tables created stores the data in the OPC Client. The system knowledge is stored in a table named Actions as shown in Figure 15. The table has three columns. The first column labelled *'TempVibOil'* is the combinations of the different fuzzy states as obtained from the field sensors. Each action is associated to a level of Priority to help prioritize the maintenance works

Figure 13. Run oil pump logic

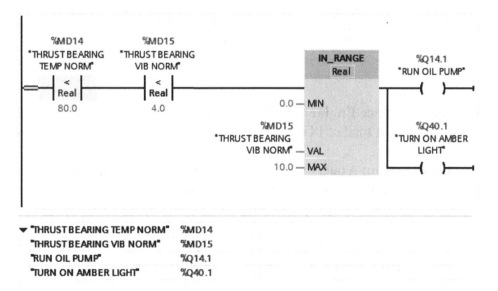

Figure 14. Temperature values in MS Access

ID	Temperature	Machine	Num	Click to Add
5/19/2014 4:34:17 PM	85	Thrust Bearing	7	
5/19/2014 4:39:07 PM	89	Thrust Bearing	8	
5/19/2014 4:39:10 PM	62	Thrust Bearing	9	
5/19/2014 4:39:12 PM	100	Thrust Bearing	10	
5/19/2014 4:39:16 PM	63	Thrust Bearing	11	
5/19/2014 5:02:19 PM	105	Thrust Bearing	12	
5/19/2014 5:02:44 PM	65	Thrust Bearing	13	
5/19/2014 5:03:24 PM	72	Thrust Bearing	14	
5/19/2014 5:05:12 PM	78	Thrust Bearing	15	
5/20/2014 12:16:13 AM	10	Thrust Bearing	16	
5/20/2014 12:17:48 AM	45	Thrust Bearing	17	
5/20/2014 1:02:08 AM	4	Thrust Bearing	18	
5/21/2014 9:52:24 AM	35	Thrust Bearing	19	
5/21/2014 12:18:39 PM	34	Thrust Bearing	20	
5/21/2014 4:02:26 PM	100	Thrust Bearing	21	
* 5/21/2014 9:37:01 PM		Thrust Bearing	(New)	

Figure 15. The Knowledge base

TempVibOil	Action	Priority
Temperature is HighVibration is HighOil Level is Low	Check bearing surface for damage or excessive clearance	Routine
Temperature is HighVibration is HighOil Level is Normal	Check machine alignment or imbalance	Priority
Temperature is HighVibration is HighOil Level is Very Low	Repair oil leak immediately, check bearing damage	Urgent
Temperature is HighVibration is NormalOil Level is Low	Investigate and topup oil at next scheduled availability	Routine
Temperature is HighVibration is NormalOil Level is Normal	Bearing temperature high,check cooling system next scheduled avail	Discretionary
Temperature is HighVibration is NormalOil Level is Very Low	Oil leak suspected, check bearing gasket	Priority
Temperature is HighVibration is Very HighOil Level is Low	Oil leak suspected, check bearing gasket and inspect bearing for dam	Urgent
Temperature is HighVibration is Very HighOil Level is Normal	Check machine alignment or imbalance imediately	Urgent
Temperature is HighVibration is Very HighOil Level is Very Low	System functionality under threat ,PLC is shutting down the unit	Emergency
Temperature is NormalVibration is HighOil Level is Low	Investigate bearing white metal pick up, topup bearing oil	Routine
Temperature is NormalVibration is HighOil Level is Normal	Bearing vibrations moderate,repair next scheduled availability	Routine
Temperature is NormalVibration is HighOil Level is Very Low	Check oil leakage and check bearing surface	Priority

4. **The Inference Engine:** Inference in Ms Access was achieved by way of queries. A total of 14 queries were designed and the code is as shown in Appendix.

5. **Fuzzification:** A query was designed, which would return a description of the level of the measured variable as *"normal, high, very high, low and very low"*. A query on "Actions" table combines the fuzzy sets and relates them to a particular rule for further processing. The query, *LastValueReading,* returns the current state of the fuzzy sets. The code of the queries is shown in Appendix below. A query which selects the recent values was designed and would return the trends on a graph in Ms Access.

Conclusion

The researchers developed the ERCM using Ms Access and S7controller. The section details the design and programming techniques applied to produce the system. The number of rules increases rapidly with the increase of the variables. In maintenance, the plant may not be too sensitive to minute and dynamic changes in operating conditions and hence some rules could be redundant and can be deleted from the system with experience while new rules are added. Combining Fuzzy systems and Expert systems is a powerful way of handling maintenance systems.

Findings and Discussions

This book discusses the results of the model designed to demonstrate the applicability of Expert systems in optimizing the operations of a hydro-power generating unit with particular attention being paid to maintenance systems. The section explains the key constituents of an Expert system which include the reliability metrics and failure mode effect and criticality analysis. The application of S7 1200 and Ms Access to code a maintenance strategy based on Expert systems was demonstrated. Maintenance decision processes involve a thorough examination and evaluation of machine status. The best decision would involve taking into consideration all or major aspects of machine life within an acceptable time duration. Different combinations of machine operating variables would result in different interpretations.

Failure Mode Effects and Criticality Analysis

The results of the Delphi research technique by way of questionnaires were processed in Ms Excel. The consensus from the experts was summarised. One round of questionnaires was used and the modal values were taken as the consensus. More rounds and follow up questioning could have improved the results. The resulting FMECA document is as shown in Appendix below. The criticality ranking is highlighted in Appendix below. Table 3 shows the criticality matrix for unit 1 power generating system which includes the turbine, generator, excitation system and the circuit breaker.

Table 3. The criticality matrix

		HIGH											LOW
		10	**9**	**8**	**7**	**6**	**5**	**4**	**3**	**2**	**1**		
OCCURRENCE	**HIGH**												
					008.4/009.4		004.1/005.0						
						001.2	007.1/009.1	007.5	004.0/005.1				
						008.1	001.1						
					003.0/007.0/007.3/009.2/009.3			001.0					
					007.2	008.5/012.0		009.5					
								010.0					
	LOW							004.2			006.0		
		10	**9**	**8**	**7**	**6**	**5**	**4**	**3**	**2**	**1**		
		HIGH					SEVERITY CLASSIFICATION				**LOW**		

Severity 9 column additional entries: 008.0 (occurrence 7), 008.3 (occurrence 5), 002.0/011.0/012.1 (occurrence 1), "\" (occurrence 9); 009.0 (occurrence 6, severity 7); 008.2 (occurrence 4, severity 8); 011.1 (occurrence 3, severity 8); 012.2 (occurrence 1, severity 8); 007.4 (occurrence 7, severity 5).

176

Criticality Analysis

- Item# 008.0 has an RPN of 54 and lies relatively on the top right hand corner of the criticality matrix. The failure mode represented by this item# is thrust bearing fatigue due to lubrication failure. The impact is relatively high as well as the probability of occurrence. This situation means that it may not be sustainable to continue running the component without taking any action, even with the condition monitoring. The redesign of the lubrication system to a more reliable one would assist to increase the availability of the thrust bearing.

- Item#008.4 represents the failure of thrust bearing cooling system and item#009.4 similarly represents the generator guide bearing cooling system. The provision of a strainer in the cooling system before the coolers would help to reduce cooler clogging rate. The scope of bearing cooler cleaning is also reduced due to pre-screening of the clogging materials. This would largely improve the return to service of the machine during a cooler cleaning outage.

- Item# 008.3 is misalignment and is of lower occurrence with high severity. Routine checking of the alignment issues, clearance and turbine blades can be carried out at each MRP.

- Item#007.5(failed gaskets is of low impact and relatively high occurrence. Gasket materials of higher quality and robustness should be employed to reduce the probability of failure.

- Item#006.0, Excitation transformer overheating is of both low probability and low impact. Though temperature monitoring is applied, 'run to failure' or 'maintenance deferred' can be applied if availability of plant is crucial, without necessarily damaging the transformer.

Reliability Metrics

1. **Calculation of MFOP:** The FMECA and the Pareto analysis done indicated the thrust bearing needed the most attention. A total of 62 inter-arrival times of the thrust bearing failure data were used to calculate the reliability data. The data was subjected to the Laplace and the Test statistic, U is 4.609 as shown in Table 4. This means that the system is repairable and is showing reliability degradation. The power Law NHPP was used to model the data.

Table 4. Laplace test statistics

Tr	118376
r	62
r-1	61
ΣTi	4840702
U	4.609457

Figure 16 shows the plot of the Expected and observed number of failures as modelled by the power law.

Applying the least squares method to Equation below, λ and δ are as shown in Table 5.

Figure 16. Modelled failure times using the Power Law NHPP

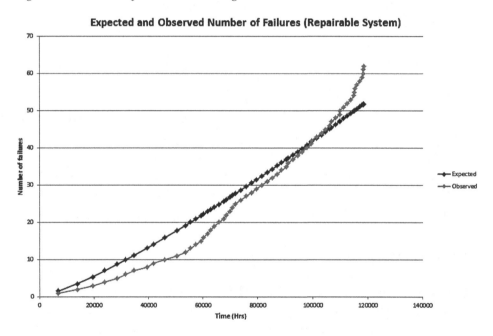

Table 5. Power Law NHPP constants

λ	2.19425E-05
δ	1.25631811

From Equation of MFOP, the MFOPS (tmf) can be obtained as in Figure 17.

From Equation, the MTBF from time t1=0 to time t2=118376hrs is calculated to be 2282 hrs. Inspection of Figure 18 shows that the MFOP equivalent of MTBF of 2282 hours has around 30% chance of survivability this clearly shows that MFOP is more realistic than MTBF. MTBF seems to leave a lot to chance given the low probability of achieving MTBF before maintenance is required. Figure 18 shows the simulated results of the developed Expert system. When temperature is *"very high"*, oil level *"very low"* and vibrations are *"very high"*, the system evaluates the machine conditions individually and as combined. It then decides to shutoff the machine because continued operation may result to damage. The PLC *writes* to the Ms Access database and at the same time sends the signal to shut off the machine by energizing the appropriate 24V dc output to energize a relay.

Figure 19 shows the condition of the machine where the vibration and temperature levels are *"normal"* while the oil level is *"low"*. The PLC start the oil pump to top up the oil as required and notifies the user about the action being taken. The user needs not to worry about the causes of the low oil level as the system has already determined that the priority is *"differed"*.

Figure 17. The probability of achieving MFOP length for the thrust bearing

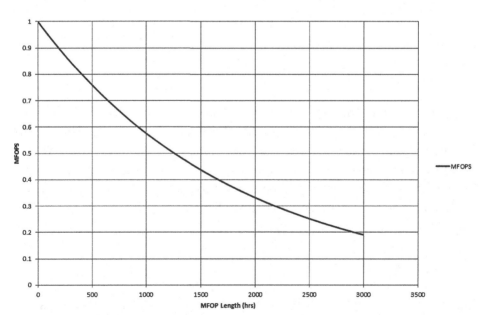

Figure 18. Mission failure imminent message from expert system

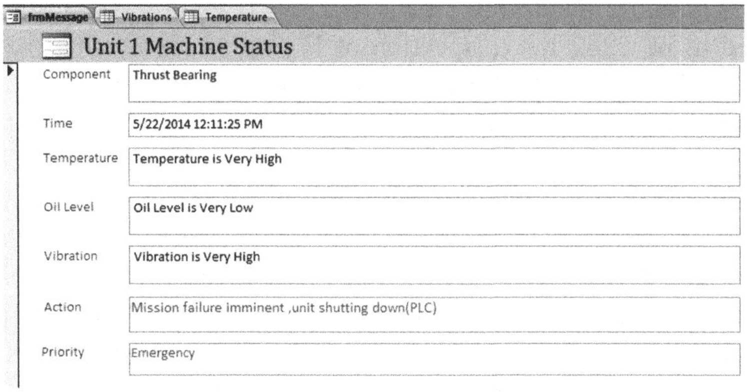

Figure 19. Oil top up message from expert system

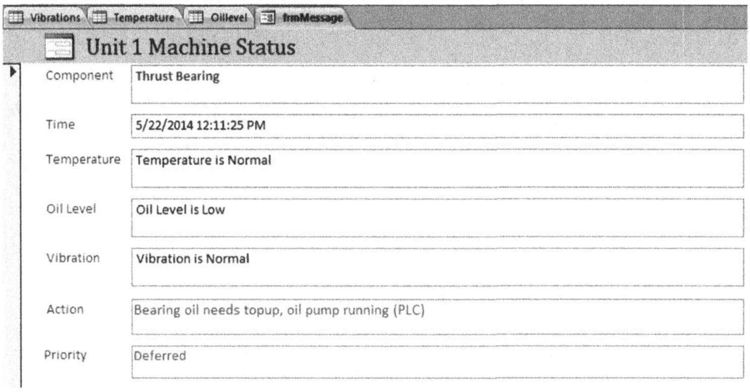

In Figure 20 all the three variables are normal. However, it can be seen from Figure 21 that the temperature trend is increasing steadily. This means that the though the actual values are ok, something could be going wrong with time and needs to be investigated. Vibrations are constant indicating a steady condition. There could be a heavy oil leak as indicated by the oil level trend. The oil was topped up to level 5 on 22 May 2014 but the oil level dropped to level 5; 15 minutes later.

Future Studies

Due to several limitations which include time, the subject under discussion was not exhausted and there is room for future studies and improvements

Figure 20. Machine running normally message from expert system

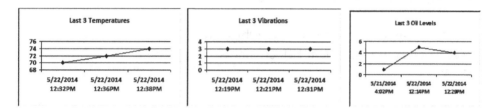

Figure 21. Temperature, vibrations, and oil level trends

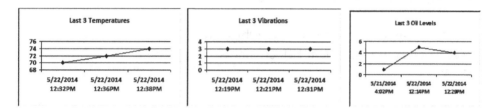

on the same topic. Reliability Centred Maintenance Expert systems can be integrated with other modern intelligent systems like the neural networks and genetic algorithms to help improve their efficiency as well as design towards autonomous maintenance systems which would carry out certain maintenance tasks.

MFOP is a relatively new reliability metric with different maintenance requirements from MTBF. Expert systems which use MFOP to plan maintenance strategies, scope and schedules can be rolled out. This means that expert systems would be integrated throughout equipment life cycle as opposed to independent application of different and in most cases non-compatible intelligent systems. Reliability analysis can be more easily and cheaply done by these systems. There is also need to consolidate experience from the workshop into a repository where expert systems can learn and get insight about the plant and prevent previous mistakes from recurring. The future of Expert RCM in developing nations should be rolled out not only in the power generation industry but across the other types of asset intensive industries.

CONCLUSION

Quick and accurate decision making in asset management in the power generation industry are key to equipment plant availability and competitiveness. Inherent human error and increasing shortage of skills demands the application of artificial and robust decision making. This dissertation has demonstrated how data can be manipulated and fused with knowledge from experts to develop a robust maintenance system based on Artificial Intelligency.

The benefits of the Expert system were seen to be its ability to substitute the human expert to make important decisions. Such a system can be trusted since it is based on the ideas of many experts, remember the Deep Blue vs. Kasparov in the 1996 Philadelphia chess tournament.

KEY TERMS AND DEFINITIONS

CPO: Competitive Performance Objectives.
OEE: Overall Equipment Effectiveness.

APPENDIX

Table 6. Thrust bearing failure data

Date	Running Time Before Maintenance (hrs)	Failure Description	Down Time (hrs)	Intervention	Outage Type (P-0/U -1)	Labour Cost/$	Lost Generation/$	Spares Costs/$	Total Costs/$
01-Apr-00	7234	Fixed time maintenance	32	Planned	0	1920	200000	1020.4	202940.4
28-Jan-01	6789	Oil leak	13	replaced leaking gasket	1	780	81250	414.54	82444.54
08-Nov-01	5678	Bearing vibrations	91	Replaced two pads	1	5460	568750	2901.78	577111.78
06-Jul-02	4407	Fixed time maintenance	29	Planned	0	1740	181250	924.74	183914.74
07-Jan-03	4576	Bearing overheat	72	Cleaned coolers	1	4320	450000	2295.91	456615.91
20-Jul-03	3067	Low oil flow	45	unblocked oil passages and replaced oil	1	2700	281250	1434.94	285384.94
26-Nov-03	3034	Dirty oil coolers	97	Cleaned coolers	1	5820	606250	3093.11	615163.11
05-Apr-04	4786	Faulty temperature recorder	0	Replaced recorder	1	23	0	0	23
21-Oct-04	2440	Fixed time maintenance	30	Planned	0	1800	187500	956.63	190256.63
01-Feb-05	4344	Oil cooler valve passing	58	Repaired valve	1	3480	362500	1849.48	367829.48
03-Aug-05	4208	Low oil level	21	Topped up oil and repaired leaking gasket	1	1260	131250	669.64	133179.64
27-Jan-06	3208	Bearing overheat	134	Cleaned strainer	1	8040	837500	4272.95	849812.95

continued on following page

Table 6. Continued

Date	Running Time Before Maintenance (hrs)	Failure Description	Down Time (hrs)	Intervention	Outage Type (P-0/U -1)	Labour Cost/$	Lost Generation/$	Spares Costs/$	Total Costs/$
15-Jun-06	1792	Dirty oil coolers	59	Cleaned oil coolers	1	3540	368750	1881.37	374171.37
31-Aug-06	1744	Leaking cooling water pipe	23	Repaired pipe	1	1380	143750	733.41	145863.41
13-Nov-06	1944	Fixed time maintenance	39	Planned	0	2340	243750	1243.62	247333.62
03-Feb-07	1008	Bearing overheat	73	Cleaned coolers	1	4380	456250	2327.8	462957.8
20-Mar-07	1504	Bearing vibrations	88	Repaired bearing	1	5280	550000	2806.12	558086.12
26-May-07	1136	Bearing vibrations	75	Changed bearing pads	1	4500	468750	2391.58	475641.58
15-Jul-07	1384	Dirty oil coolers	45	Cleaned coolers	1	2700	281250	1434.94	285384.94
13-Sep-07	1688	Faulty oil level switch	5	changed oil level switch	0	300	31250	159.43	31709.43
22-Nov-07	1616	Bearing overheat	44	Repaired cooling water valve	1	2640	275000	1403.06	279043.06
30-Jan-08	1064	Fixed time maintenance	24	Planned	0	1440	150000	765.3	152205.3
16-Mar-08	1184	Bearing vibrations	100	Smoothed bearing surface	1	6000	625000	3188.77	634188.77
08-May-08	1045	Bearing overheat	51	Cleaned coolers	1	3060	318750	1626.27	323436.27
23-Jun-08	1003	Water valve leaking by the gland packing	5	Repacked gland	1	300	31250	159.43	31709.43

continued on following page

184

Table 6. Continued

Date	Running Time Before Maintenance (hrs)	Failure Description	Down Time (hrs)	Intervention	Outage Type (P-0/U -1)	Labour Cost/$	Lost Generation/$	Spares Costs/$	Total Costs/$
04-Aug-08	1998	Bearing vibrations	9	Topped up oil	1	540	56250	286.98	57076.98
26-Oct-08	1944	Oil contamination with water	8	Replaced oil	0	480	50000	255.1	50735.1
16-Jan-09	1889	Bearing overheat	43	Cleaned coolers	1	2580	268750	1371.17	272701.17
06-Apr-09	1884	White metal pick up	89	Repaired bearing	1	5340	556250	2838.01	564428.01
28-Jun-09	1980	Excessive clearance	38	Adjusted pads	0	2280	237500	1211.73	240991.73
20-Sep-09	1860	Bearing running noisy	14	Adjusted bearing housing	1	840	87500	446.42	88786.42
07-Dec-09	1752	Faulty thermocouple	2	Replaced thermocouple	1	120	12500	63.77	12683.77
18-Feb-10	1874	Low water flow to coolers	10	Cleaned strainer	1	600	62500	318.87	63418.87
07-May-10	1622	Fixed time maintenance	43	Planned	1	2580	268750	1371.17	272701.17
16-Jul-10	1798	Low oil level	15	replaced leaking gasket	1	900	93750	478.31	95128.31
29-Sep-10	700	Bearing overheat	62	Cleaned coolers	1	3720	387500	1977.04	393197.04
31-Oct-10	1528	Bearing inspection	3	Inspection	0	180	18750	95.66	19025.66

continued on following page

Table 6. Continued

Date	Running Time Before Maintenance (hrs)	Failure Description	Down Time (hrs)	Intervention	Outage Type (P-0/U -1)	Labour Cost/$	Lost Generation/$	Spares Costs/$	Total Costs/$
03-Jan-11	1672	Bearing vibrations	340	Repaired cracked runner	1	26400	2750000	14030.61	2790430.6
28-Mar-11	1608	Oil contamination with water	32	Replaced oil	1	1920	200000	1020.4	202940.4
04-Jun-11	1784	Bearing overheat	14	Repaired lube pump	1	840	87500	446.42	88786.42
18-Aug-11	1644	Low water flow to coolers	75	Cleaned coolers	1	4500	468750	2391.58	475641.58
28-Oct-11	167	Bearing vibrations	196	Repaired bearing	1	11760	1225000	6250	1243010
13-Nov-11	1674	Faulty water flow relay	29	Replaced water flow relay	0	1740	181250	924.74	183914.74
23-Jan-12	1692	Bearing damage	147	Replaced bearing	0	8820	918750	4687.5	932257.5
08-Apr-12	1534	Excessive wear particles in oil	35	Re-metaled bearing	0	2100	218750	1116.07	221966.07
13-Jun-12	1487	Bearing overheat	167	Replaced bearing	1	10020	1043750	5325.25	1059095.3
20-Aug-12	634	Fixed time maintenance	32	Planned	1	1920	200000	1020.4	202940.4
17-Sep-12	1536	Low oil level	23	replaced leaking gasket	1	1380	143750	733.41	145863.41
21-Nov-12	1332	Bearing overheat	72	Cleaned coolers	1	4320	450000	2295.91	456615.91
19-Jan-13	168	Bearing inspection	2	Planned	0	120	12500	63.77	12683.77

continued on following page

186

Table 6. Continued

Date	Running Time Before Maintenance (hrs)	Failure Description	Down Time (hrs)	Intervention	Outage Type (P-0/U -1)	Labour Cost/$	Lost Generation/$	Spares Costs/$	Total Costs/$
26-Jan-13	1344	Bearing vibrations	9	Topped up oil	1	540	56250	286.98	57076.98
23-Mar-13	1367	Oil contamination with water	61	repaired cooler and Replaced oil	1	3660	381250	1945.15	386855.15
22-May-13	1278	Fixed time maintenance	35	Planned	0	2100	218750	1116.07	221966.07
15-Jul-13	1115	Bearing overheat	15	Cleaned coolers	1	900	93750	478.31	95128.31
31-Aug-13	224	Faulty oil level switch	1	Repaired switch	1	60	6250	31.88	6341.88
10-Sep-13	133	Bearing running noisy	4	realigned bearing housing	1	240	25000	127.55	25367.55
15-Sep-13	867	Bearing vibrations	188	Repaired bearing	1	11280	1175000	5994.89	1192274.9
29-Oct-13	1001	Bearing overheat	102	Topped up oil	1	6120	637500	3252.55	646872.55
14-Dec-13	987	Dirty oil coolers	19	Cleaned coolers	1	1140	118750	605.86	120495.86
25-Jan-14	57	Faulty oil level switch	3	changed oil level switch	1	180	18750	95.66	19025.66
28-Jan-14	213	Bearing vibrations	260	Re-metaled bearing	1	15600	1625000	8290.81	1648890.8
17-Feb-14	146	Fixed time maintenance	20	Planned	0	1200	125000	637.75	126837.75
Total	**118376**		**3470**		**62**	**214223**	**22312500**	**113838.97**	**22640562**

Chapter 8
Programmable Logic Controller Using Fuzzy Logic for Water Gate Re-Design and Condition Monitoring for Dams

ABSTRACT

The research on gate control work presents the concept of regulating the flow of water by controlling the dam gates or shutters using servo-motors in order to manage the dam water level. Water level control and insufficient water supply are the major challenges at the hydropower company (HDPC). The system at this dam site consists of the outlet works tower with four outlet penstocks that collects water to the main discharge outlet. The automation implements the electronic control system that uses the programmable logic controller (PLC) and the optical level sensors to detect water level. For the new design, the power screw hoist mechanism is used to open and close the gate of weight 600N with the efficiency of 35.4% of the square thread screw. The gate takes 160 seconds to travel a vertical distance of 600mm in its guide, to fully open and close the 400mm penstock diameter. Mat-lab Simulink was used to control the electronic system for stability to avoid vibrations and fluctuation of speed.

DOI: 10.4018/978-1-5225-3244-6.ch008

LAYOUT OF THE TOWER OUTLET WORKS AND THE GATES

The system used to open and close the gates is a rope and chain mechanism which is operated manually by people on duty according to water level and supply of water needed. Figure 1 it shows the layout of the tower which is at the deepest point of the dam.

The tower consists of four penstocks inside it that collect water according to its level to the main discharge outlet away from the dam. It is made up of non-permissible concrete which does not allow water to sip into the tower. The flat vertical gates are used to close the water at each penstock, working under its own weight in the wheel guides as shown in Figure 2. In the tower are the ladders used to go down to each and every level where the gate and penstocks are located as shown in Figure 3. There are four level stages which are accessible using this ladder, each penstock has a butterfly valve to open and close the water.

Water from the collecting tower is then piped to the irrigation and to the mine. The water that is supplied is raw water at low charging rates. 2000m³ is supplied to the fields for irrigating per month whilst 1000m³ is supplied to the mine for the processes.

Figure 1. Collection outlet tower at WDC dam

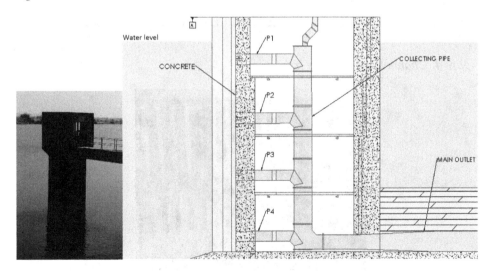

Figure 2. Rope and chain manual mechanism at Eben dam

Figure 3. Access ladder down the tower.

CHALLENGES FACED AT WDC

The system has failed to supply sufficient water to the irrigation and the mine due to the manual system use in opening and closing the gates. When water level changes there is a difficulty is open the gate that corresponds to that level since a person will be required to open the gates at the moment it

Figure 4. Channelling water to irrigation

changes to avoid cutting of water supply. Oversupplying the water has also been encounter due to gates not fully closed by $400m^3$ and $350m^3$ per month to the irrigation and mine respectively. It takes time to open the gates at another level using the ladder. In the case of floods this system would not be able to act as an effective measure of floods. Over supplying water means a loss is also encountered. Gates fail and crush sometimes.

GENERATION OF DESIGN CONCEPTS

Introduction

The following concepts were generated from the standard specifications and however, paying due regard to copyright violations. Not all of the details are shown in design drawings. It provides with the simple sketches as the idea is to bring out the concept only without comprising on clarity. The notes are also given to as they describe how the design concept operates. The concept selection will be considered using suitable criteria to come up with the most suitable design. The generated concepts are going to be chosen depending on the most important parameters such as it functionality, efficiency, quality, reliability and economic cost.

Concepts

The gate is attached to the wire rope which enables the gate to move upward and downwards in its guides. The rope passes over the wheel to the pulley where it is wound around it, the pulley is driven by the shaft coupled from the gearbox. The gearbox reduces the speed of the motor to the required speed. The motor and the gearbox are coupled in between them by a shaft. The rotation of the motor is the one that provides the movement of the gate up and down according to water level through a reduction gearbox.

Advantages of the Rope Drum Mechanism

- Close under their own weight (Sehgal & Ala, 1987).
- Wire rope can maintain the gate in any given position without fear of drifting (Sehgal & Ala, 1987).
- Familiar technology

Figure 5. Design Concept 1 2D diagram
Courtesy of Solidworks15 by T. Mushiri.

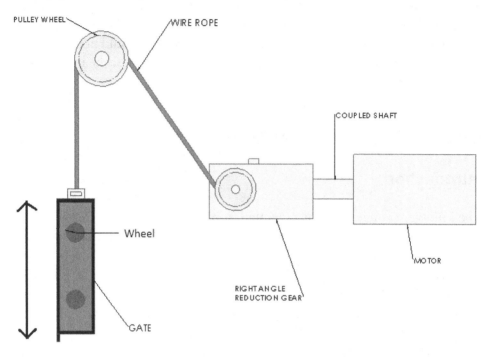

Figure 6. Concept 1 3D diagram
Courtesy of solidworks15 by T Mushiri.

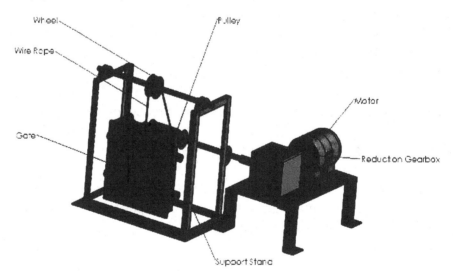

Disadvantages of the Rope Drum Mechanism

• Restrict their use to the fixed wheel gate.
• Less sufficient force to effectively press the bottom seal to prevent leakage.
• Rope is subjected to wear and may have to be frequently replaced.
• Stretching of the wire rope can cause problems with the actuation of the limit switches causing difficulty in operation.

The motor drives the reduction gearbox through a shaft while the gearbox also drives the smaller chain sprocket which is attached to bigger chain sprocket by a chain. The driven sprocket rotates with the pulley drum on the same shaft. The driven sprocket and the drum pulley rotates in either direction moving the gate upwards and downwards in its concrete guide according to water level.

Disadvantages of the Rope and Chain Mechanism

• Failure of a limit switch or the solenoid brake can cause serious accidents (Sehgal & Ala, 1987).
• Flaws in the length adjustment of several wire rope or chain on the gate can prove costly (Sehgal & Ala, 1987).

Figure 7. Design concept 2 2D diagram
Courtesy of Solidworks15 by T. Mushiri.

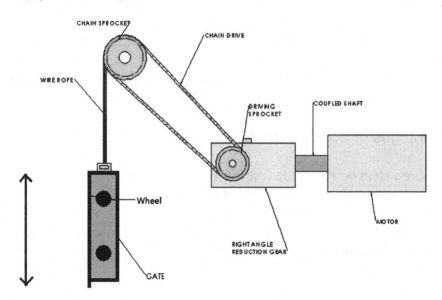

Figure 8. Concept 2 3D diagram
Courtesy of solidworks15 by T Mushiri.

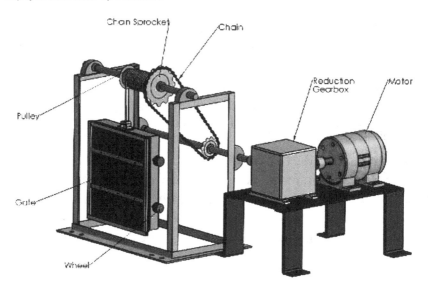

- Closing the gate by gravity during electrical power failure cannot be provided without providing expensive additional equipment (fan brake) (Sehgal & Ala, 1987).
- Design costs are high and inspection must be frequently (Sehgal & Ala, 1987)

The sluice gate consists of four rollers/wheels that enable the gate to move up and down in its guides according to the water level. The gate is attached to a stem (long threaded shaft) and it is a rising stem thus the threaded shaft does not rotate but rather moves up and down as the lifting nut attached to the gate frame rotates. The lifting nut is driven by the chain sprockets assembly which is also driven by the reduction gearbox coupled to the motor.

Advantages of Screw Stem Mechanism

- Screw stem hoist provides the positive force both for up and downward movements of the gate (Sehgal & Ala, 1987).
- Less expensive in terms of repairing intervals.
- It ensures complete seal of the gate when closed (no water leakage).
- Long life span and proves to be simple.

Figure 9. Design concept 3 2D diagram
Courtesy of Solidworks15 by T. Mushiri.

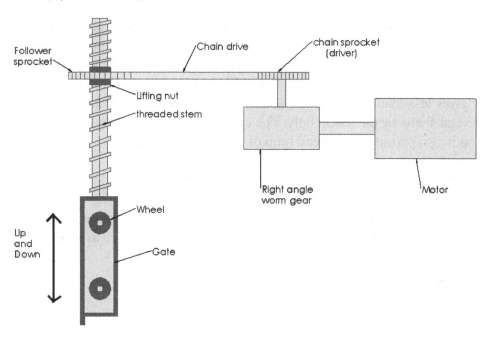

Figure 10. Concept 3 3D diagram
Courtesy of solidworks15 by T. Mushiri.

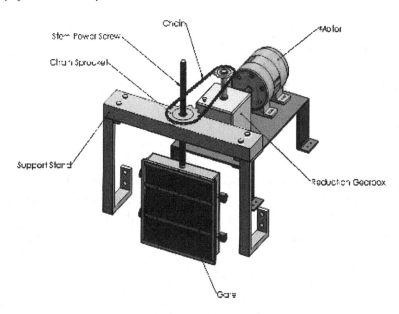

Advantages of Chain Drives

There is no slip takes place during chain drive, hence perfect velocity ratio is obtained and the chains are made of metal, therefore they occupy less space in width than a belt or rope drive. It may be used for both long as well as short distances providing a high transmission efficiency (up to 98 percent). It gives less load on the shafts and it has the ability to transmit motion to several shafts by one chain only. The chain transmits more power than belts and it also permits high speed ratio of 8 to 10 in one step.

Disadvantages of Screw Stem Mechanism

Minor misalignment between the stem and the lifting nut can cause rapid wear (Sehgal & Ala, 1987). These mechanisms are normally applied to large gates.

Design Concept Selection

The three concepts generated will be evaluated using the following parameters below;

A: Function
B: Reliability
C: Ease of maintenance
D: Efficiency
E: Easy of manufacturability
F: Simplicity of layout
G: Ergonomics
H: Weight
I: Product lifespan
J: Quality
K: Cost

The two selected criteria which are the binary dominance matrix and the criteria weighting are used. Binary dominance matrix compares two parameters against each other and the most considered parameter is chosen and a mark is awarded to it.

Each criterion has been assigned a weight that ranges from 1-10. The criteria are multiplied by the weight from the binary dominance matrix to give the concept weight.

From the Table 2 the concepts were weighed. Concept 1 have the weighting of 418, concept 2 the weighting of 346 and concept 3 the weighting of 527. The concept with more weighting is chosen. Concept 3 was chosen as the best since it has 527 > 418 and > 346.

Conclusion

Using the binary dominance matrix and the weighting criteria from the table above concept 3 is indicated as the best concept. The choice of concept three was chosen based on the parameters that will enable the system to function efficiently, reliable and at the same time the system should be easy to maintain. Moreover, the system should be economical (cheap compared to other mechanisms) and the price should not limit or cause under design of the system operation and its efficiency. The next chapter outline the detailed design of the major components of the system.

Table 1. Binary dominance matrix at Eben dam

	A	B	C	D	E	F	G	H	I	J	K	Total
A	\	1	1	0	1	1	1	1	0	1	1	8
B	0	\	1	0	1	1	1	1	0	1	0	6
C	1	1	\	0	1	0	1	1	0	0	0	5
D	1	1	1	\	1	1	1	1	1	0	0	8
E	0	1	0	0	\	0	1	1	1	1	0	5
F	1	1	1	0	0	\	0	1	1	1	0	6
G	0	0	0	0	1	0	\	1	1	1	1	5
H	0	1	0	0	1	1	0	\	1	1	0	5
I	1	1	1	1	0	0	0	0	\	1	0	5
J	0	0	1	1	0	0	0	0	1	\	1	4
K	1	1	1	1	1	1	0	0	1	0	\	7

Table 2. Weighting criteria

	Weight	Concepts			Concepts Weight		
		1	**2**	**3**	**1**	**2**	**3**
A	8	7	6	9	56	48	72
B	6	6	6	9	36	36	54
C	5	7	5	8	35	25	40
D	8	7	6	10	56	48	80
E	5	5	4	7	25	20	35
F	6	5	5	8	30	30	48
G	5	6	5	7	30	25	35
H	5	7	6	5	35	30	25
I	5	7	6	7	35	30	35
J	4	6	7	10	24	28	40
K	7	8	6	9	56	42	63
	Total				418	346	527

PROGRAMMABLE LOGIC CONTROLLER USING FUZZY LOGIC FOR WATER GATE RE-DESIGN AND CONDITION MONITORING FOR DAMS

Introduction

A full detailed design of the chosen concept 3 of the power screw mechanism is done in this chapter outlining all the major components to be designed. The selection of suitable material is going to be done for every component as per specifications. The detailed design consists of the following components:

1. Flat vertical gate
2. Power screw
3. Chain drives
4. Shafts and its keys
5. Sizing of the motor and the reduction gear box.

The threaded stem moves the gate up and down as it is the rising stem (non-rotational) while the lifting nut rotates. The lifting nut is seated on top of the roller thrust bearing and the thrust washer as shown by the sectioned view in Figure 11. The motor and the gearbox acts as the actuators of the

Figure 11. Cross section of the power screw mechanism (lifting mechanism)
Courtesy of Solidworks15 by T. Mushiri.

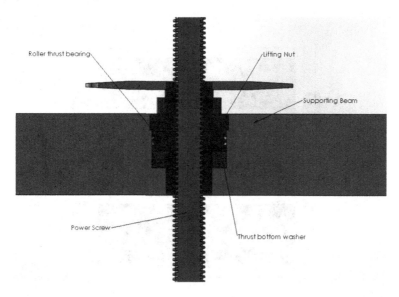

system. The motor is coupled to the speed reduction gearbox (Right angle worm gear) which drives the chain drives allowing rotation of the lifting nut. The rotation of the lifting nut allows the stem to move up and down depending on the direction of which the motor is rotating.

Material Selection of a Flat Vertical Gate as Per Specification

The gate consists of four wheels that will enable the gate to move vertically up and down in the concrete guides. The material for the wheels is Cast steel as specified by IS NO 1030, 1998.The front face is coated thin layer of rubber (PVC) to prevent direct corrosion (pitting) of the steel sheet of the gate. The skin thickness should 9.5mm including the 1,5mm to cater for corrosion as specified by IS 4622. It also consists of rubber (PVC) seal that prevent water leakages, the resilient rubber seal shall be mounted on the slide and shall be held securely in place with a retainer bar bolted to the side. The gate is reinforced with the High water resistance stainless steel to prevent rusting. The wheel axles are made of Chrome nickel steel which is corrosion resistance steel according to IS 2004, 1994.

Figure 12. Sluice gate
Courtesy of Solidworks15 by T. Mushiri.

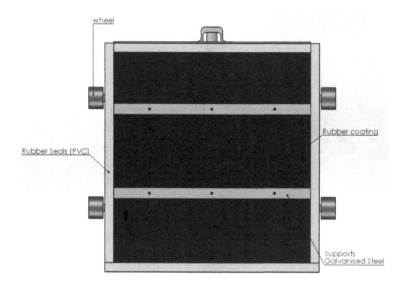

Force Acting on the Vertical Gate

Figure 13 shows a view of how the gate is laid out in the tower. The design considers the force acting on the gate closing the fourth penstock (the bottom penstock where greater pressure is exerted).

Considering a vertical area submerged below the surface of the water as shown above.

$$TotalForce(F_t) = \rho g B \frac{(y_2^2 - y_1^2)}{2} \tag{1}$$

but

$$\frac{y_2 + y_1}{2} \tag{2}$$

is the distance from the free surface to the centroid y and

$$y_2 - y_1 = D \tag{3}$$

Figure 13. View of the gate closing the penstock diameter
Courtesy of solidworks15 by T. Mushiri.

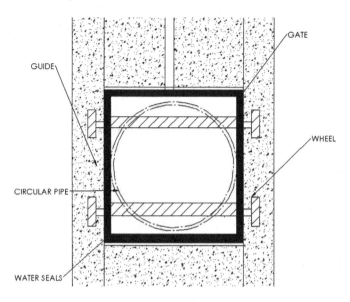

$$F_t = \rho g A y \qquad (4)$$

(F_t) acting on the gate due to water pressure, where ρ=density g=force due to gravity A= area of the gate at which the pressure is acting.

Taking $\rho = 10^3 kg / m^3$ g=9.81 m / s² length*width of the gate = (0.61×0.60) m

The height of the water level from the centre of the bottom diameter of the pipe to the surface is 40.5m

$$(F_t) = 10×9.81×0.61×0.60×40.5$$

=145413.63N is the total force exerted on the gate

The design specification of the gate is that; it should not exceed 60kgs When the mass of the gate is 60 kg then its weight is given by;

$$W = mg \qquad (5)$$

Weight of the gate = mg = (60×9.81) =588.6 N

Force due to friction on the wheel and the concrete guide

$$F = \mu N \tag{6}$$

where $\mu = 0.33 \, friction \; coefficient$ between the steel wheels and the concrete and N is the normal force action perpendicular to the gate face.

F= 0.33×145413.63 = 47986.29N

In this case the frictional force due to seals is neglected (assumption). Design of the power screw is shown in Figures 14 and 15.

Design Considerations

The length of the stem for which the threads maybe provided shall be the sum of the following:

1. The total lift of the gate.
2. Length of the nut in contact with the stem.

Figure 14. Stem rod and lifting nut
Courtesy of Solidworks15 by T. Mushiri.

Figure 15. Sectioned view of the power screw
Courtesy of Solidworks15 by T. Mushiri.

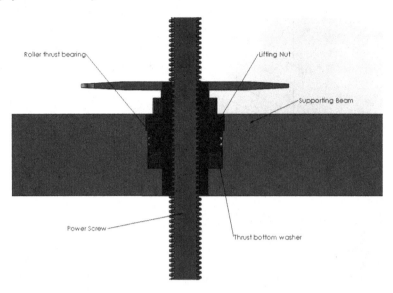

3.   Extra allowance of 300mm

The stem should be able to provide power transmission in either direction. The square threads are chosen over acme and buttress threads, the material used in the stem screw is steel. Figure 16 shows the square threads profile.

1.   Assumptions Considered in Design of Stem and Lifting Nut:
   a.   Frictional resistance induced by the pressure of the water against the gate is neglected.
   b.   Frictional resistance caused by the seals is neglected.
   c.   Weight of the stem rod is neglected.

Torque Required to Raise and Lower the Gate Using Square Threaded Screw

Considering the following parameters and the outside diameter and pitch from Appendix $d_o = 50mm$, Pitch = 8mm $\mu_1 = 0.12$ *for the washer*, R_1 and R_2 of the washer are 50mm and 100mm respectively R=37.5mm

$$\mu = Tan\,\phi \qquad (7)$$

Figure 16. Square threads
Courtesy of solidworks15 by T. Mushiri.

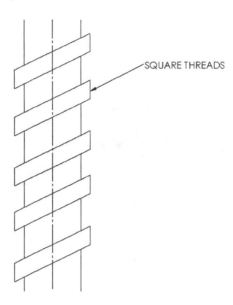

which is 0.1, the coefficient of friction between the screw and nut. The inner diameter or core diameter of the screw is given by,

$$d_c = d_0 - p ,$$
(8)

Core diameter of the screw =50-8= 42mm

$$d = \frac{d_0 + d_c}{2}$$
(9)

Mean diameter of the screw = 46mm

$$\text{Tan}\,\alpha = \frac{p}{\pi d} \frac{8}{\pi \times 46} = 0.05535$$
(10)

Considering a Case of Raising the Gate

Since the frictional resistance acting in the opposite direction to the motion of screw is neglected.

Total load acting on the screw $W = 588.6\text{N}$

And the torque required to overcome the friction at the stem screw

$$T_1 = P * \frac{d}{2}\left[W * \text{Tan}(\alpha + \varphi)\right] = W\left\{\frac{\text{Tan}\,\alpha + \text{Tan}\,\varphi}{1 - \text{Tan}\,\alpha\,\text{Tan}\,\varphi}\right\}, \quad T_1 = 2114.8\text{Nmm}$$
$$= 588.6\left(\frac{0.05535 + 0.1}{1 - 0.05535 \times 0.1}\right)\frac{466}{2}$$

$$(11)$$

Mean radius of the washer,

$$R = \frac{R_1 + R_2}{2} \tag{12}$$

and the torque to overcome friction at the washer is given by,

$$T_2 = \mu W R;\; T_2 = 0.12 \times 588.6 \times 37.5 = 2648.7\text{Nmm} \tag{13}$$

Total torque required to overcome friction

$$T = T_1 + T_2 = 2114.8 + 2648.7;\; T = 4763.5\text{Nmm} \tag{14}$$

In the Case of Lowering the Gate

The frictional resistance will act upwards when lowering the gate but in this case it is assumed to zero

Total load acting on the screw $W = 588.6\text{N}$

Torque from equation 11 gives $T_1 = 601.14$Nmm

Torque require to overcome friction from equation $13 = 0.12 * 588.6 * 37.5$, $T_2 = 2648.7$Nmm

Total torque required to overcome friction $T = T_1 + T_2 = 3249.84$Nmm

Efficiency of the Square Threads

Since it is assumed that there is no frictional resistance between the screw and the nut, the angle of friction becomes zero. The efficiency of the square threads becomes;
Efficiency,

$$\eta = \frac{IdealEffort}{ActualEffort} = \frac{W \tan\alpha}{W \tan(\alpha + \varphi)} = \frac{0.05535}{0.156215} \tag{15}$$

Efficiency of the Square threads is 0.354 or 35%

Number of Threads and Height of the Nut

The material used on the lifting nut is Cast iron or bronze while on the screw is Steel. From Appendix (Table 17) on hoisting Screw the bearing pressure is 4.2N/mm^2 and t = 4mm
The bearing pressure,

$$P = \frac{W}{\pi d t n} \tag{16}$$

where n is the number of threads in contact with the nut, H is the height of the lifting nut and t is the thickness of the threads.

$$n = \frac{29.12}{8} = 3.61,$$

therefore the number of threads is 4

$$H = np = 4 \times 8 = 32 \tag{17}$$

From the Appendix (Table 16) we consider the height of the nut as 40mm
Length of the stem is the sum of 600mm required to raise and lower the gate and 40 mm in contact with the stem and the extra allowance of 300 mm as per specification.
Total stem length is 940mm.

Stresses in the Screw Stem

Maximum compressive stress in the screw is given by

$$\sigma_c = \frac{W}{A_c} = \frac{W}{(\pi/4)(d_c)^2} = \frac{588.6}{1385.55} = 0.4248N \ / \ mm^2 \tag{18}$$

Maximum shear stress in the threads,

$$\tau = \frac{16T}{\pi d_c^3} = \frac{16 \times 4763.5}{\pi \times (42)^3} \tag{19}$$

Maximum shear stress in the threads is 0.3274N/mm².
Therefore, the maximum shear stress in the threads is given by

$$\tau_{max} = \frac{1}{2}\sqrt{\sigma_c^2 + 4\tau^2} = 0.5 \times \left[\left(0.4248^2 + 4 \times 0.3274^2\right)\right]^2 = 0.39N \ / \ mm^2 \tag{20}$$

Power Required to Drive the Lifting Nut

$$P = T\omega \tag{21}$$

The cutting speed of the hoisting screw from the Appendix (Table 17) is 6-12m/min and considering 6m/min
And the speed of the screw is given by

$$N = \frac{CUTTING\ SPEED}{PITCH} = \frac{6000}{8}$$

$$Power = 4.7635 \times \frac{2\pi \times \dfrac{6000}{8}}{60}$$

The power required to drive the lifting nut is = 374.12 W or 0.374kW. The Table 3 shows the calculated parameters of the stem and the nut.

Sizing of the Motor and Gearbox Specifications

The motor should be able to transmit the required power and torque by the lifting nut. It should also be able to overcome all the frictional between the gate and the gate guides, forces due to weight of the gate and internal friction in the reduction gearbox. The motor should be able to rotate in both directions to enable the movement of the screw up and down, a medium starting torque is required. It should be a 3 phase ac motor as per IS325 and insulated according to class B of IS1271. The parameters to be considered on the selection of the motor are as follows:

1. Power of half horse power (0.5hp or 0.37kw)
2. Torque of 4.763kNm.
3. Application of uniform load, shock and vibration.
4. Running hours per day and suitable to work under wet conditions (corrosive areas).

Table 3. Power screw designed parameters

Description	Parameter
Square thread pitch	8mm
Number of threads in contact with the lifting nut	4
Height of the lifting nut in contact	40mm
Efficiency of the square thread power screw	0.35
Power required to by the lifting nut	0.374KW
Diameter of the stem (outside, core and mean) respectively	50,42,46
Length of the stem including 300mm of allowance	940mm

3 phase AC induction motors are classified in two that is the squirrel cage and slip ring induction motor. In this case the Squirrel cage type is going to be used because of the advantages over the slip ring type shown in Table 4.

Courtesy of the specifications of electric motors.

The motor torque should be always higher than the load torque for the load to be accelerated to the rated power. Graph (b) in Figure 17 shows the correct selection of the motor where $C_p > C_r$.

Specifications of the chosen electric motor are shown in the Table 5 from the TEC Motor Technical catalogue. The motor selected in the catalogue is the Squirrel cage induction motor which uses Ac current in 3 phase with 0.5 Horse power at the speed of 1750 rpm.

Table 4. Advantages of squirrel cage and slip ring type

Type	Squirrel Cage Induction Motor	Slip Ring Induction Motor
Design	Squirrel cage rotor	Wound rotor
Starting current	High	Low
Starting torque	Moderate	High
Starting /rated current	High	Low
Breaking torque	>160% of the rated torque	>160% of the rated torque
Efficiency	High	Moderate
Protection device	Simple	Simple
Required space	Small	Large space
Maintenance	Small	For slip rings and brushes
Cost	Low	High
Wet environment. Explosive areas.	Very good (operate good)	Not recommended for wet conditions

Figure 17. (a) Incorrect (b) correct graphs of starting torque and resistive load where C_r = the load is torque (resistive load) and C_p = the motor torque (starting torque).

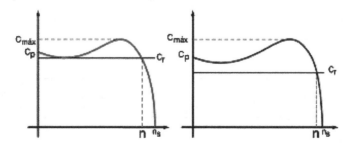

Table 5. Specifications of the electric motor

Ratings	0.37KW (0.5hp) at 1750 rpm 50hz 3ph
Power Factor	0.74
Voltage	230V or 400V
Full load Current	1.90A at 230V, 1.11A at 400V
Efficiency	65%
Physical Properties	Weight 6kg 132mm wide ×255mm long ×188mm high

Parameters considered in selection of the reduction gearbox are as follows:

1. Input speed 1750 rpm
2. Target output speed 30 rpm (at gear ratio)
3. Service factor
4. Orientation.
5. The gearbox should be designed for 15% more than the torque of the screw/lifting nut 5KNm.

Since the gearbox will be operating after a period of time (not 24 hours) and driving a moderate load the service factor of 1.25 is considered from the appendix. Service factor is multiplied by (K) = 1 since it is driven by the electric motor or hydraulic motor and the class of service for this gear is II. For the target output speed of 30 rpm the 60:1 gear ratio is selected from the appendix (table 18) to give the output speed of 29 rpm. The orientation of the gearbox should be right –angle worm gearbox. The efficiency of a gear reducer 60 of size 718-15 is 0.90 shown in the appendix (Table 19) the specifications are shown in Table 6.

Table 6. Gear reducer specifications

Type	Right Angle Worm Gear
Output speed	29 rpm
Input speed	1750rpm
Gear reducer	60:1
Size	718-15
Efficiency	0.90

Figure 18. Roller thrust ball bearing
T Mushiri; Solidworks, 2015.

Consideration Made in Selection of Roller Contact Bearing

The speed of the selected bearing would be a very low speed of 15 rpm thus it has less speed to that of the speed specified (high speeds). The load on the bearing is moderate compared to heavy loads. This system will be used for a short periods and its breakdown would not have serious consequence (Khumi & Gupta, 2005). Material to be used should be according to IS NO 305:1998 which is phosphor bronze, self-lubricating and of high strength brass castings. Therefore, the roller contact thrust bearing was selected (roller thrust ball bearing) size 51212 with a bore of 60mm and 95mm outside diameter. The capacity of the roller thrust bearing is 68KN with the reliability of 90% as shown in Figure 19. The total load on the bearing is 600N at a speed of 14.5 rpm that give the life of the bearing in hours or in revolution. The life of the bearing is 1456622.112652×10 revs or 1674278290.404765 hours according to the solidworks15 bearing calculator.

The frame should be able to support all the weight exerted by the gate and the stem all in total being 600N. The material should be water resistance and resistance to corrosion since it will be working under wet conditions. It should be coated or painted to reduce corrosion and pitting. The legs should be able to withstand the load of the gate and that of the top support beam. Stainless steel/cast carbon steel is used on the support since it can be used in a corrosive environment, high strength to mass and no rusting which mean no paint is necessary. It has a wet abrasion resistance and it can also prevent catalytic reaction due to contamination (no contamination).

Figure 19. Capacity of bearing and its life span
Solid works, 2015.

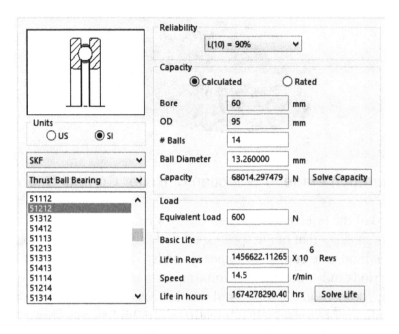

Stresses and Deflection of the Top Support Beam

The material used is cast carbon steel with the yield strength of 248.17MPa. Using the safety factor of 5 for the material

$$\sigma_y = \frac{248.17}{5} = 49.63 Mpa$$

Figure 20. Support frame of the gate and power screw
Courtesy of solidworks15 by T Mushiri.

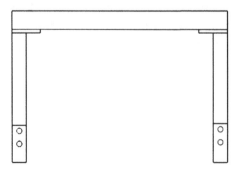

Using the beam calculator in solid-works the principal stresses in the x and y axis are 8.955KPa and 40.268456MPa respectively as shown in appendix.

$$\sigma_x = 8.955Kpa$$
$$\sigma_y = 40.268456Mpa$$

Calculating the von Mises using the equation

$$\sigma_{vm} = \sqrt{\sigma_x^2 + \sigma_y^2 - \sigma_x\sigma_y} \prec \sigma_y \quad ;$$
$$\sigma_{vm} = \sqrt{8955^2 + 40268456^2 - 8955 \times 40268456} = 40.26Mpa \quad (22)$$

Thus the von misses $\sigma_{vm} < \sigma_y$, that is 40.26Mpa < 49.63Mpa therefore the design is safe.

The deflection of a simple supported beam carrying a load of 600N is 0.8855mm shown by the beam calculator in appendix which is less than the recommended 1.5mm.

The power to be transmitted is 0,374kW with the driving sprocket speed 29 rpm from the gearbox and the follower sprocket speed it 14.5 rpm. The following requirements must be met in the design of a chain drive:

1. The sprocket teeth should have sufficient strength so that they will not fail under static loading or dynamic loading during normal running conditions.
2. The sprocket teeth should have wear characteristics so that their life is satisfactory.
3. The use of space and material should be economical.

Figure 21. Chain drives
Courtesy of solidworks15 by T. Mushiri.

4. The alignment of the sprockets and deflections of the shafts must be considered because they have effect on the performance of the chain drive.

Design procedure of chain drives is given:

Given data; Rated power = 0.5 HP (0.37KW), $N_1 = 29 rpm$, $N_2 = 14.5 rpm$

1. The velocity ratio,

$$V \cdot R = \frac{N_1}{N_2}; 29/14.5; \text{Velocity ratio} = 2 \tag{23}$$

2. Selection of the number of tooth on the smaller sprocket, T_1

    The number of teeth on the smaller sprocket plays an important role in deciding the performance of a chain drive. A small number of teeth tends to make the drive noisy. A large number of teeth makes chain pitch smaller which is favourable for keeping the drive silent and reducing shock, centrifugal force as well as the frictional force (Khumi & Gupta, 2005).
    Hence from appendix (Table 20), the number of teeth on smaller sprocket is 27 using roller chain. $T_1 = 27$
    The number of teeth on the larger sprocket,

$$V \cdot R = \frac{T_2}{T_1} = 2 \times 27, \ T_2 = 54 \tag{24}$$

3. Determination of the design power by using the service factor, such that Design power = Rated power × Service factor.

$$DP = RP * K_s \tag{25}$$

Determination of the service factor, Ks of this system.

$$K_s = K_1 * K_2 * K_3 \tag{26}$$

1. System will be working under constant load, load factor (K) = 1, for constant load.
2. System will be lubricated over a certain period of times, lubrication factor = 1.5, for periodic lubrication.
3. Rating factor = 1.5, for continuous service.

Design power =$0.374 \times 2.25 = 0.8415 kW$

From the appendix the design power of 0.8415KW corresponds to chain type number 08B per strand. Therefore the roller chain with a single strand can be used to transmit the require power. We find that the Pitch, p =12.7 from appendix.

Roller diameter, d= 8.51 mm;

Minimum width of roller, b= 7.75 mm;

Breaking load, WB=17.8 kN

Pitch Circle Diameters of the Driver Sprocket and the Follower Sprocket

The pitch circle diameter of the smaller sprocket or pinion (gearbox),

$$d_1 = p \cos ec\left(\frac{180}{T_1}\right) d_1 = 12.7 \times 8.6137 = 109.39mm \qquad (27)$$

Table 7. Service factor values

Parameters	Values
K_1	1
K_2	1.5
K_3	1.5
$K_S = K_1 \times K_2 \times K_3$	2.25

Pitch circle diameter of the larger sprocket (follower),

$$d_2 = p \cos ec \left(\frac{180}{T_2} \right); \; d_2 = 12.7 \times 3.33 = 218.4mm \tag{28}$$

Pitch line velocity of the smaller sprocket,

$$V_1 = \frac{\pi d_1 N_1}{60} \tag{29}$$

$$V_1 = \pi \times 0.05287 = 0.1661m \, / \, s$$

Therefore the load on the chain

$$W = \frac{RatedPower}{PitchLineVelocity} \frac{0.374}{0.1661} = 2.252kN \tag{30}$$

$$Factorofsafety = \frac{W_B}{W}; \; \frac{17.8}{2.252} = 7.90 \tag{31}$$

This value is more than the value given in appendix (Table 21) in the Appendices, which is equal to 7. Considering the minimum centre distances between the range of 30-50 times between the bigger and smaller sprocket. Taking it as 35 times the pitch. Therefore,

$$\textit{The centre distance between the sprockets} = 35p \; 35 \times 12.7 = 444.5mm \tag{32}$$

In order to accommodate initial sag in the chain, the value of centre distance is reduced by 2 to 5 mm. Therefore correct centre distance

$$x = 444.5 - 4 = 439.5mm \tag{33}$$

K is the number of strands of the chain,

$$K = \frac{T_1 + T_2}{2} + \frac{2x}{p} + \left[\frac{T_2 - T_1}{2\pi} \right]^2 \frac{p}{x}$$

(34)

K=40.5+69.21+0.53359=110.2

therefore the $K = 111$ chain linkages
    Therefore the length of the chain; $L = Kp$

L=111*12.7=1409.7mm / 1.4097m

The List of Design Material Requirements

The motor speed is 1750 rpm connected to the right angle worm gear of 60:1 speed reduction. The chain drive also provides speed reduction of 2:1 to the driven sprocket that drives the lifting nut. The overall speed reduction of the system is 120:1 from the motor. The rotational speed of the lifting nut is 14.5 rpm that enable the stem to move up and down. The stem moves 8mm in one revolution of the lifting nut. Considering the safety factor of 1.3 between the rotation of the steel screw and the bronze nut.
    The stem will travel at

$$14.5 \, rev \, / \, min \times 8mm \, / \, rev \times 1.3 = 150.8 \text{mm/min}$$

Since the diameter of the penstock pipe is 400mm it takes 2.65 minutes (2minutes and 40sec) to fully open the full diameter pipe.

Table 8. List of chain specifications

Chain type	08B roller chain
Chain length × 1 strands	1.41m
Pitch circle diameters	
Gearbox (pinoin) sprocket (d_1), $T_1 = 27$	109.4 mm
Stem (follower) sprocket (d_2), $T_2 = 54$	218.4 mm
Centre distance, **X**	439.5mm

Figure 22. Chain drive system

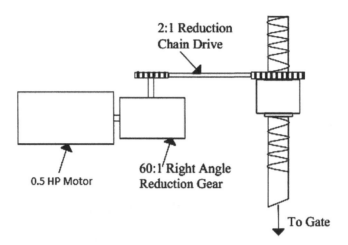

Design of Shafts

The shaft is the transmission type that transmit power from the electric motor to the reduction gear box subjected to twisting. Carbon steel / alloy steel can be used in this case. The carbon steel has good mechanical properties 40 C 8 with the ultimate tensile strength of 560 – 670 Mpa and yield strength of 320 MPa. It has high strength, high wear resistance, good machinability and good heat treatment properties.

The power transmitted is 0.374Kw at the speed of 1750 rpm. The ultimate shear stress of 360 MPa with a safety factor of 8.

$$\tau = \frac{360}{8} = 45MPa. \tag{35}$$

Power transmitted by the shaft is

$$P = \frac{2\pi NT}{60} \tag{36}$$

Torque transmitted (T) $= \dfrac{60 \times 374}{2\pi \times 1750}$ =20.4Nm / 20400Nmm

The torque of 2.04 Nm is equivalent to the equation

$$T = \frac{\pi * \tau * d^3}{16};$$
(37)

The diameter of the shaft is given by $d = \sqrt[3]{\left(\frac{16 \times 20400}{\pi \times 45}\right)}$ =36mm / say

40mm standard size

Design of the Keys and Muff Coupling

Diameter of the shaft (d) = 40mm, Yield strength for the shaft =360 MPa for carbon steel.

From appendix the standard size of the key width (w) = 14 mm, thickness (t) = 9 mm. (parallel rectangular key).

Material for the key is steel with yield strength 320 Mpa. FS = 2

Using maximum shear stress theory, the maximum shear stress of the shaft is

$$\tau_{max} = \frac{360}{2 * 2} = 90 MPa$$
(38)

Maximum shear stress for the key $\tau_k = \frac{320}{2 * 2} = 80 MPa$. Torque to be

transmitted by key and the shaft is $T = \frac{\pi \times 90 \times 40^3}{16} = 1.131 \times 10^6$ Nmm (From

equation 37). Considering the failure of the key as a result of shearing when the torque transmitted is

$$1.131 \times 10^6 Nmm = l \times \tau_k \times w \times \frac{d}{2} = l \times 80 \times 14 \times 20$$

$$l = 50.49 mm$$

Failure of the key due to crushing when the torque is

$$1.131 \times 10^6 Nmm = l \times \frac{t}{2} \times \sigma_{ck} \times \frac{d}{2} = l \times \frac{9}{2} \times \frac{320}{2} \times 20$$

$$l = 78.54 mm \text{ of the key}$$

Therefore, the larger value is considered, thus 78.54mm say 80mm
The shaft strength factor according to H.F Moore is,

$$e = 1 - 0.2\left(\frac{w}{d}\right) - 1.1\left(\frac{t}{2d}\right) = 1 - 0.2 \times 0.35 - 1.1 \times 0.1125 = 0.806 \ ; \tag{39}$$

Shaft Coupling

The shaft coupling is used to provide connection between the motor and the gearbox and also ease of disconnection of the two parts. The shock loads transmitted are reduced from one shaft to another, at the same time protecting against overloading. The Rigid coupling is used in this case since there is need for perfect alignment between two shafts. Sleeve or muff coupling is suitable.

1.   The shaft diameter is 40mm from Equation of the shaft above.
2.   Design of the sleeve.

Cast iron used on the muff assuming the maximum allowable shear stress of 15Mpa
Outside diameter of the muff

$$D = 2d + 13mm = 2 \times 40 + 13 = 93mm \tag{40}$$

And the length of the muff is $l = 3.5d = 3.5 \times 40$

$l = 140mm$.

Now considering the induced shear stress in the muff τ_c of cast iron .The muff is a hollow shaft transmitting maximum torque given by;

$$1.131 * 10^6 \, Nmm = \frac{\pi * \tau_c * (D^4 - d^4)}{16D} 1.131 \times 10^6 = \frac{\pi \times \tau_c \times \left(93^4 - 40^4\right)}{16 * 93}$$
$$= 1.525 \times 10^5 \tau_c, \tau_c = 7.41N \, / \, mm^2$$

$$\tag{41}$$

The induced stress in the cast iron muff is 7.41Mpa which is less than the permissible shear stress of 15 MPa. Therefore, the muff is safe.

The width and thickness of the key is 14 mm and 9 mm respectively. The length of the key in the shaft is given by;

$$l = \frac{3.5d}{2} = 70mm$$

The maximum allowable shear and crushing of steel key are 80Mpa and 185Mpa respectively.

Checking for induced crushing and shear in the key, we consider the shearing of the key.

$$\text{Torque (T)} = 1.131 \times 10^6 \, Nmm = l \times w \times \tau_s \times \frac{d}{2} = 70 \times 14 \times \tau_s \times 20$$

$$1.131 \times 10^6 \, Nmm = 1.96 \times 10^4 \tau_s \, ; \, \tau_s = 57.7 Mpa$$

Now checking for crushing of the key, we have

$$1.131 \times 10^6 \, Nmm = l \times \frac{t}{2} \times \sigma_{cs} \times \frac{d}{2} \, ; \, = 70 \times 4.5 \times \sigma_{cs} \times 20 \, ;$$

$$1.131 \times 10^6 \, Nmm = 6.3 \times 10^3 \sigma_{cs} \, ; \, \sigma_{cs} = 179.5 Mpa$$

Since the induced shear and crushing of the key are less than the permissible stresses therefore the design of the key is safe. The design parameters of the muff are shown in Table 9.

Stress Analysis and Simulation of Mechanical Components

The stress analysis of the following components is done using solidworks2015;

1. Gate
2. Top beam support
3. Bearing
4. Leg support for the beam
5. Lifting

Table 9. Design parameters of the muff

Rigid Coupling Type	Sleeve or Muff Coupling
Length (L)	140mm
Diameter of muff (D)	93mm
Diameter of the shaft (d)	40mm
Length of the key	70mm

The designed gate is made up of AISI 4130 Steel, normalized at 870C/ Stainless steel, it has to withstand a force of 145.41363kN exerted by water on its face.

The exerted stresses on the face of the gate are in the range below $2.126 \times 10^8 N / m^2$, which is less than the yield strength of the selected material. Much of the exerted force is on the wheel axles which is observed to be safe with a value of $3.188 \times 10^8 N / m^2$ as shown in figure above. Stress analysis of the top beam support is shown in Figure 25.

The beam support carries the whole load of 600N of the gate and the stem. The lifting beam should have a very small deflection of less than 2mm due to the load applied, this is so to prevent the damage of the gate and the motor

Figure 23. Gate stresses analysis (von Mises)
Courtesy of solidworks15 by T Mushiri.

*For a more accurate representation see the electronic version.

Figure 24. Back view of the gate (von Mises)
Courtesy of solidworks15 by T Mushiri.

For a more accurate representation see the electronic version.

Figure 25. Beam stress analysis (von Mises)
Courtesy of solidworks15 by T Mushiri.

For a more accurate representation see the electronic version.

since if it deflects more it shortens the travelling distance than required causing overdriving of the gate. The material used on the beam is alloy steel as an alternative of cast carbon steel and stainless steel. The load carried by the beam exert the von Mises stress that are in a range of $7.725 \times 10^5 N / m^2$, which is less than the maximum stress that can be reached by the beam at the maximum load of $1.030 \times 10^6 N / m^2$. The beam is safe since the maximum

Figure 26. Bearing stress analysis (von Mises)
Courtesy of solidworks15by T Mushiri.

Model name:bearing
Study name:Static 2(-SKF - 51212- 14 DE,AC,14_68-)
Plot type: Static nodal stress Stress1

von Mises (N/m^2)

3.381e+006
3.099e+006
2.818e+006
2.536e+006
2.255e+006
1.973e+006
1.691e+006
1.410e+006
1.128e+006
8.467e+005
5.651e+005
2.835e+005
1.948e+003

→ Yield strength: 1.930e+008

For a more accurate representation see the electronic version.

stress does not exceed the yield strength of $4.60 \times 10^8 \ N \ / \ m^2$ shown in Figure 26. The deflection of this beam when a 600N load is applied is 0.00109582 mm.

The selected roller thrust bearing can withstand the 600N exerted on it, with a nearly.

$1.410 \times 10^6 \ N \ / \ m^2$ less than yield strength of $1.93 \times 10^8 \ N \ / \ m^2$ as shown in the Figure 17. The suitable material to use on selection of the bearing is Phosphor bronze 10% D, UNS C52400.

The legs that support the beam are able to withstand the stress exerted on them, the material used is AISI 4130 Steel, normalized at 870°C. The range of the stresses are between $1.517 \times 10^6 \ N \ / \ m^2$ and $1.760 \times 10^6 \ N \ / \ m^2$ which are less than the maximum stress of $3.033 \times 10^6 \ N \ / \ m^2$, therefore the design of the support legs is safe since it less than the yield strength of $4.60 \times 10^8 \ N \ / \ m^2$ as shown in Figure 27. The lifting nut is made up of 4130 Steel, normalized at 870C, as an alternative of cast iron and bronze. It should be able to withstand all the stresses acting on it caused by the stem rod carrying the gate load of 600N. The stresses are in a range of $6.244 \times 10^5 \ N \ / \ m^2$ which is less than the maximum yield strength of $4.60 \times 10^8 \ N \ / \ m^2$ as shown in Figure 28.

The major components are analysed for stresses, to see if it fails. The most critical parameter that has to be monitored is the displacement to avoid

Figure 27. Stress analysis of the legs and displacement
Courtesy of solidworks15 by T Mushiri.

For a more accurate representation see the electronic version.

Figure 28. Stress analysis of the lifting nut (von Mises)
Courtesy of solidworks15 by T Mushiri.

For a more accurate representation see the electronic version.

damaging the actuators and the gate. However, it is seen that all components designed can withstand the forces and the stresses, the designed components are safe.

Design of Electronic Control System

The following will be looked at in this section of the control system.

1. Selection of sensors.
2. How the gates in the tower operates.
3. Block Diagram of the whole system.
4. Ladder logic diagram.
5. Stability of the system.

   The type of sensors to be selected should be a point level sensors that can detect the level of water above or below a set point. The following application parameters should be considered on selection of the sensors high precision, accuracy, response rate and ease of calibration or programing. The sensor should be capable of being immersed in water. The types of point level sensors are conductive and optical. The optical sensor out class the conductive sensors because of the two electrodes in contact with water corrode. There are frequently replaced and require expensive material for electrodes (titanium or gold). An optical sensor is used because there are reliable and dependently since there are not affected by vapour and can be easily removed and cleaned.

   The sensors are immersed in water in this case thus the continuous light from the Infrared LED is refracted out from the prism to the water, no light is reflected back to the receiver. When the tip of the prism no longer sensing water (contact of water) the light is reflected back to the receiver giving the feedback that the water level is below the sensor.

   The sensors are arranged in such a way that there at midpoint of the penstock diameter to accommodate the time interval of opening each gate which is 160 seconds to fully open thus (continuous supply of water is required). There are four sensors that senses water level with each penstock having one sensor as shown on Figure 29.

1. When the water level is rising:
    a. When water level is at the marked point (sensor 1) the gate 3 opens.
    b. When water level is at the marked point (sensor 2) the gate 2 opens while gate number 3 closes.
    c. When water level is at the marked point (Sensor 3) the gate 1 is opened and gate 2 closes.
2. When water level is lowering (draining):
    a. When water level is above sensor 3 the gate 1 is open.

Figure 29. Layout of Sensors immersed in water
Courtesy of solidworks15 by T Mushiri.

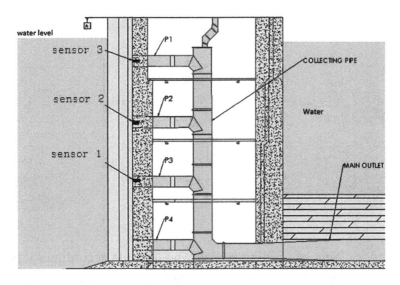

b.   When water level is at point sensor 3 the gate 2 opens and gate 1 closes.

c.   When water level is at point sensor 2 the gate 3 opens and gate 2 closes.

Ladder Logic Diagram

Ladder logic represents how the gates will be operating as the water level rises and lowers.

- **Limit Switch:** Enable the motor or the gate not to over travel that is to avoid damage the gate and motor.
- **On Delay Timer:** Serves the same purpose as the limit switch since it will only delay before the set time is reached, when it reaches the set time of 160s to fully open and close the gate it will stop.

Key:

- **T01:** On delay timer.
- **Motor Forward:** Opening the gate
- **Motor Reverse:** Closing the gate.

Figure 30. Block diagram of the system

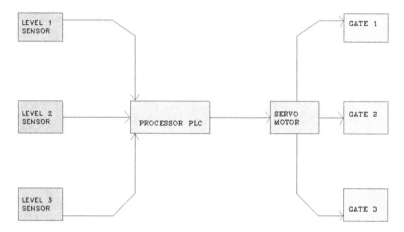

The ladder logic of opening and closing the gate according to the water level is shown on the SMT (IMO PLC) software sheet, and simulations of the ladder was done successfully. It is shown in appendix for PLCs.

FUZZY LOGIC STABILITY OF SYSTEM LAYOUT OF WATER

Overall Transfer Function of the System

The mathematical model of the system from the system block diagram is as follows, considering the armature circuit the overall transfer function is;

$$G(s) = \frac{\theta_m(s)}{v_b(s)} = \frac{K_t}{S\left[(R_a + SL_a)(J_m S + D_m) + K_b K_t\right]} \tag{42}$$

Table 10. Ladder logic inputs (symbols)

Sensor 1	I01
Sensor 2	I02
Sensor 3	I03
Limit switch 1	m04
Limit switch 2	m03
Limit switch 3	m06

Table 11. Ladder logic outputs

Motor 1 coil (forward)	M01
Motor 2 coil (forward)	M02
Motor 3 coil (forward)	M05
Motor 1 coil (Reverse)	M07
Motor 2 coil (Reverse)	M08
Motor 3 coil (Reverse)	M0D

Figure 31. System block diagram
Courtesy of MatLab15 by T. Mushiri.

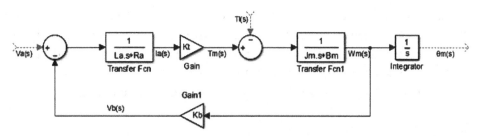

Therefore, substituting the above parameter, the transfer function from equation 42 becomes;

$$G = \frac{0.01}{S^4 - 0.1169S^3 + 0.001687S^2 - 9.0081 \times 10^{-7}S}$$

Table 12. Input parameters of the motor

Torque constant K_t	0.01
Armature inductance L_a	0.1
Armature impedance R_a	1
Co-efficient of friction D_m	1
Back e.m.f armature constant K_b	11
Total inertia of the motor J_m	0.0112

The system response to the step input without a controller is shown in Figures 32 and 33.

The scope above shows output response of the input step function; it shows that the system will overshoot before reaching its steady state value of 6.3. The system has a large overshoot that cause the system to be unstable resulting in vibration, and damaging the components of the system.

System Modification and Control

The system is now simulated with a controller which is the (PID controller) which is tuned until the system is stable shown in Figure 34. The PID compensate the system to be stable preventing unwanted vibrations and disturbances. The output response of the controlled system is shown in Figure 35. The system has risen during the delay time of about 40second before reaching the desired output (steady state value).

The tuned information of the Controller is shown in appendix.

Figure 32. Simulink model for the System assuming the system is an LTI system

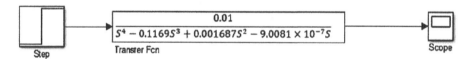

Figure 33. Output response of the uncontrolled system
Courtesy of Mat Lab, 2015 by T Mushiri.

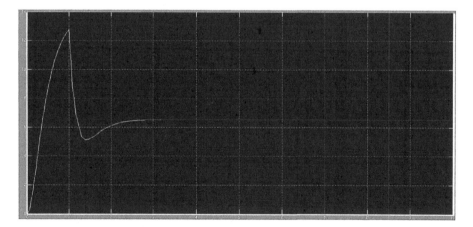

Figure 34. Modification of the system with a PID controller

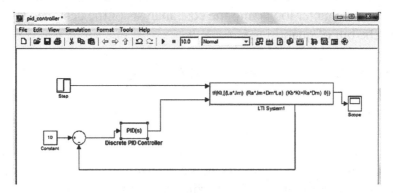

Figure 35. Output response of the controlled system (Stable)
Courtesy of Matlab Simulink, T. Mushiri.

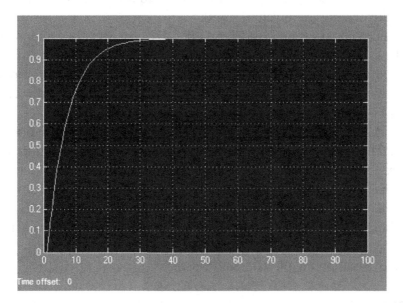

Conclusion

The system with an uncontrolled input function will be unstable and with overshoots. In this system, an uncontrolled signal will result in vibrations, fluctuations in speed. The result will be increased rates of wear and tear of the machinery components. The advantages of the controlled system are that it can increase machine uptime and higher production. Zero instability reduces tear and wear of rotating components and thus reduced system maintenance cost. The next chapter looks into the cost analysis of the designed system by creating the bill of quantities.

Cost Analysis of the Designed Concept

Introduction

The purpose of this section is to undertake economic analyses of the design to compare the costs with benefits and to determine if this has an acceptable return. The cost and benefits of the project must therefore be identified once they are known they should be priced and their economic values determined. All the material used are locally available. The BOQ of both the mechanical design and the electronic design are constructed below in Tables 13 and 14. All the dimensions are given in (mm).

The mechanical system cost US $2065.5, the cost of every component was from the local market. The price of the components that need to be machined is incorporated and included in the BOQ. The electronic system components will be selected depending on which type of PLC to be used. In this case the IMO SMT client brand is used.

The SMT IMO PLC gadgets cost $3098.50 including installation of the system. The quotation of the prices was done using the global market to compare the prices. The total cost of the whole system is $ US5164.00, this system has a high cost to implement it but in the long run it will pay back the initial cost. This system is reliable since it also reduces the maintenance cost, labor, and gate replacement. Water saving are achieved when using this system because there is assurance that all gate are closed at the required time prevent over supplying of water and wastages.

Payback Period

2000m^3 and 1000m^3 per month is supplied to Eben irrigation scheme and Madziwa mine respectively. An amount of 2400m^3 and 1350m^3 of water is recorded to be supplied to these designated areas. 750m^3 of water are wasted (oversupplying) due to improper closing of gates by people. The system will be fully automated thus it also reduces the human labour to monitor the gates. Currently four people are required to be operating the system which

Table 13. Bill of quantities (price of locally available market)

Part #	Description of Part	Quantity	Material	Cost/Item	Total Cost
1	Right angle worm reduction gearbox (size 718-15) Gear reducer (60:1)	4		$65.00	$260.00
2	Squirrel cage induction motor 3ph (0.5 horse power)	4		$55.00	$220.00
3	08B roller chain (1 strand ×1410mm)	4	Stainless steel	$4.50	$18.00
4	Chain sprocket (pinion) D_1=109.5mm	4	Alloy steel	$8.00	$32.00
"	Chain sprocket (follower) D_2=218.5mm	4	Alloy steel	$8.00	$32.00
5	Lifting nut	4	Bronze/cast iron	$20.00	$80.00
6	Square threaded stem/screw L=940mm pitch=8mm	4	Mild steel	$60.00	$240.00
7	Roller thrust bearing (size 51212) d_i=60 d_o=95	4	phosphor bronze	$5.00	$20.00
8	Thrust washer (d_i=60 d_o=95)	4	Cast iron	$1.00	$4.00
9	Motor-gearbox shaft (40mm×150mm)	8	Carbon steel/alloy	$7.50	$60.00
10	Muff coupling (L=140mm D=93)	4	Cast iron	$10.00	$40.00
11	Muff keys (L=80mm, 1×1mm)	4	Steel	$1.00	$4.00
12	Shaft keys (L=70mm, 1×1mm)	4	Steel	$1.00	$4.00
13	Flat sheet (600×1000)	4	Steel/cast iron	$10.00	$40.00
14	Gearbox-motor stand legs (150×300) (3.5 Kgs)	16	Cast iron	$6.00	$96.00
15	Adjustable bolts M15 +Nuts	16	Mild steel	$1.00	$16.00
16	Bolts M15 + Nuts	16	Mild steel	$0.90	$14.40
17	Bolts M30 + Nuts	32	Mild steel	$1.00	$32.00
18	Support frame legs (80×300×720)mm (3.5Kgs)	8	Stainless steel / carbon steel	$10.50	$84.00
19	Top support of frame beam(100×200×1200)mm (10Kgs)	4	Carbon steel	$30.00	$120.00
Gate Break Down Parts					
1	Steel sheet (face) (10×600×600)mm (5kgs)	8	Steel	$5.00	$40.00
2	Side sheet (100×600) (4kgs)	8	Steel	$4.00	$32.00
3	Top & bottom sheets(100×600)mm (4kgs)	8	Steel	$4.00	$32.00
4	L-seal support bars (25×20×600) 2 at the bottom and 2 at the side (1kg)	16	Stainless steel	$2.00	$32.00
5	Reinforcement inside (100×550) (2-per gate) (4kgs)	8	Stainless steel	$4.00	$32.00
6	Cross flat steel sheet holding the coating (5×30×550)mm	8	Stainless steel	$1.00	$8.00
7	Wheels (d=60mm w=40) (2.5kgs)/4	16	Cast iron	$2.50	$40.00
8	Rubber seals (15×25×600)mm + (15×25×550) mm	16	Pvc /natural rubber	$0.90	$17.10
9	Axles (20×70)mm	16	Chrome nickel steel	$1.00	$16.00
	Assembly of the system	4		$100.00	$400.00
	Cost of mechanical system	4			**$2065.50**

Table 14. Table bill of quantities (electric gadgets)

Part #	Description of Part	Quantity	Material	Cost/Item	Total Cost
1	Optic Sensors (LV171 wet probe)	3	Stainless steel	$60.00	$180.00
2	Sensor cables (5kgs)	3	Fibre	$20.00	$60.00
3	Limit switches	3	Steel	$17.5	$52.50
4	Miniature circuit breaker	1		$20.00	$20.00
5	SCADA software (IMO SMT client)	1		$600.00	$600.00
6	PLC data logger software	1		$300.00	$300.00
7	PLC controller	1		$200.00	$200.00
8	On delay timer	1		$48.50	$48.50
9	Motor coils	14		$3.50	$49.00
10	PID controller	1		$30.00	$30.00
11	Emergency push button	1		$29.00	$29.00
12	Communication cable from laptop to PLC	1		$30.00	$30.00
13	IMO PLC (s7-300) front connector	1		$500.00	$500.00
	Cost of electrical gadgets				**2098.50**
	Cost of Installation				**$1000.00**
	Total cost of PLC				**3098.50**

is currently available, with this system only two people will be monitoring the system. $400 per month is considered as the allowance for each person. The payback period to be meet should be five years. The Table 15 shows the total saving of implementing the new system

$$PaybackPeriod = \frac{TotalInitialAmountInvested}{AnnuallyOperationSaving} = \frac{5164}{1250} = 4.13 \qquad (43)$$

Table 15. Table annual operating cash flow amount

	Water Saving (m³)	Cost per m³	Saving per Month	Annual Savings
Madziwa Mine	400	$ 0.05	$ 20	$ 240
Eben irrigation Scheme	350	$ 0.05	$ 17.5	$ 210
	Firing	**Saving**		
Labour	2	$ 400		$ 800
Total Savings				**$ 1250**

The payback period with the annual saving of $ 1250 is 4 years and 2 months. The whole system costed $ 5164, with the annual saving of $1250 giving the payback period of 4 years and 2months which is less than the expected payback period of 5years. It is a reasonable period implementing such a system with a high initial cost. This section outlines the proper procedures and recommendations to be done on the design system (power screw mechanism). Further studies need to be done to improve the efficiency of the system of the power screw which is 35% efficient on the design since there is room to improve the efficiency up to its maximum (re-designing). The material selection is also to be looked at for every component and change be made depending on the rate of wearing of material due to rust, loads and frictional resistances. This section will give an insight of every major component of the system on how there are maintained and least but not last, the investment of the system.

Maintenance and Installation of the Mechanical System That Will Improve the Breakdowns

The installation of the electric motor should be in a way that allows easy of access to inspection and maintenance. Adequate protection should be considered since the environment will be wet and dust at times. It should not be covered by a box that restricts free air circulation. The base of the motor should be bolted on the flat surface to avoid free vibrations. It can be mounted on concrete surface or metallic stand as shown below by figure below.

- **Maintenance:** The motor should be kept clean form the dust, debris and oil. Soft materials should be used to clean the motor e.g brushes, cotton. It should also be lubricated in different intervals using appropriate grease and checking for proper alignment of the shafts.
- **Bearings:** Proper alignment should be done on installation to prevent damage of the bearings and vibration caused by misalignment. They should be lubricated in a time interval bases as preventative and corrective maintenance. Failure of the bearing may cause a big damage to the lifting nut and the stem threads. Water proof bearing are required. The bearing should be able to withstand the thrust and radial loads.
- **Reduction Gearbox:** The shaft connection and alignment should be checked in four places around the shaft at 90°. The correct gap must also be maintained and the gear drive must be level and secure

before alignment of shafting begins (Cleveland, 1999). A quality lubricating grease is recommended for all coupling applications. A sufficient amount of clean lubricant is required for long service life (Cleveland, 1999). It should be checked for leaks on the seals and lose bolts. Tooth profile should be checked periodically to ensure maximum transmission.

- **Lifting Nut and Stem Rod:** The lifting nut should be properly aligned to the stem rod prevent damage of the threads. The used on the design is bronze or cast iron. Selection of another material would be necessary if the lifting nut wore out fast. In order to achieve maximum transmission it should be changed regularly and lubrication should be done each time maintenance is done.

- **Sluice Gate:** The gate does not need periodic maintenance, but visual inspection of crack and wearing out due to corrosion and pitting. Rust may cause weak points on the gate which cause water sipping through. The steel sheet must be cleaned and painted to avoid rusting. The groove where the gate seals seat must be cleaned to make sure that there is positive thrust and complete water tight.

Optical Level Sensors

1. **Installation of Level Sensor:** The water level sensor should be installed vertical or horizontal depending on which point of level is to be detected, it should be suspended in water and fixed to the walls by clamps to protect it from the external forces (Undue forces) as shown in Figure 36. If it is to be installed where there is a water flow it should be installed with a sensor stand.

2. **Maintenance of the Level Sensor:** Less maintenance is require because of the optical connection terminals and the cable. A routine check-up important to prevent inaccuracy and troubleshooting. Checking for corrosion on the terminals and damage of the cable is important. The sensor casing should be checked for corrosion. The optical connector and the terminals should be checked if there are tightened correct as shown in Figure 37.

System Stability

The system with an uncontrolled input function will be unstable and with overshoots. In this plant, an uncontrolled signal will result in vibrations,

Figure 36. Installation of the level sensor in the water
Courtesy of paint by T Mushiri, 2016.

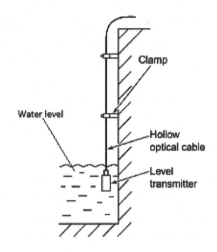

Figure 37. Proper connection of the connector
Courtesy of paint by T Mushiri, 2016.

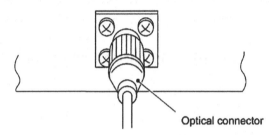

fluctuations in speed. The result will be increased rates of wear and tear of the machinery components. Thus the system should always be stable to increase its uptime and prevent damage of components. Figure 37 shows how the system should be when it is in a stable state.

1. **Working Environment of the PLC (Whole Electronic System):** Since the control system will be installed and working in the area with moisture, all the major components should be built in an enclosed control box. When the system fails, such that it cannot open and close the gates there is a backup plan that uses the manual way of opening the penstocks. It consists of the butterfly valves at each penstock that will be used to open and close for water.

CONCLUSION

All the objectives were met and the design specifications were satisfied as per outlet works diagram provided from WDC. Two methods, binary dominance matrix and the weighting criteria were used to select the best concept. The designed system uses a stem or power screw to open and close the gates. The motor, reduction gearbox and the chain drives acts as the actuators. The control system uses the servomechanism, with the PLC. The optic level sensors are used to sense water level then give the feedback to the processor for opening and closing the dam shutters. This designed system serves to be efficient and reliable but however some modifications needs to be done to improve the efficiency. Material selection also need to be chosen so as to improve reduce the rate of corrosion and pitting. This automated control system serves to eliminate all the challenges and human disabilities (errors) which were encountered by using the manual system at the WDC that is, ensuring proper water distribution or supply to the community and irrigation scheme and water level control will be managed. The proposed system is reliable since no human effort will be made to open and close the gate according to water level, it also monitors the dam's safety. This system does not only monitors the vibration on the dam structure caused by excessive water levels but can be implemented in irrigation schemes and act as an effective measure of floods without destroying the gates.

REFERENCES

Cleveland, O. (1999, November). *Enclosed Gear Drive Maintenance Manual.* Retrieved April 14, 2016, from http://www.horsburgh-scott.com/resources/PDFs/hs-maint-manual.pd

Khumi, R. S., & Gupta, J. K. (2005). *Machine Design.* Academic Press.

Sehgal, C. K., & Ala, F. G. (1987). *Operation and maintenance of hoisting equipment for flood gates for locks and dams introduction* (2nd ed.). American Society of Mechanical Engineers.

APPENDIX

Table 16. Basic dimensions for square threads in mm (Normal series) according to IS: 4694 – 1968 (Reaffirmed 1996)

d_1	d	D	d_c	p	h	H	A_c
(46)	46	46.5	38				1134
48	48	48.5	40	8	4	4.25	1257
50	50	50.5	42				1385
52	52	52.5	44				1521
55	55	55.5	46				1662
(58)	58	58.5	49	9	4.5	5.25	1886
(60)	60	60.5	51				2043
(62)	62	62.5	53				2206
65	65	65.5	55				2376
(68)	68	68.5	58	10	5	5.25	2642
70	70	70.5	60				2827
(72)	72	72.5	62				3019
75	75	75.5	65				3318
(78)	78	78.5	68				3632
80	80	80.5	70				3848
(82)	82	82.5	72				4072

Table 17. Limiting values of bearing pressures

Application of screw	Material		Safe bearing pressure in N/mm²	Rubbing speed at thread pitch diameter
	Screw	Nut		
1. Hand press	Steel	Bronze	17.5 - 24.5	Low speed, well lubricated
2. Screw jack	Steel	Cast iron	12.6 – 17.5	Low speed < 2.4 m / min
	Steel	Bronze	11.2 – 17.5	Low speed < 3 m / min
3. Hoisting screw	Steel	Cast iron	4.2 – 7.0	Medium speed 6 – 12 m / min
	Steel	Bronze	5.6 – 9.8	Medium speed 6 – 12 m / min
4. Lead screw	Steel	Bronze	1.05 – 1.7	High speed > 15 m / min

Table 18. Class II service, gear set with single reduction (service factor 1.25)

Reducer ratio	Output rpm	Input horsepower at 1,750 rpm													
		7	8	9	10	11	12	13	14	15	16	17	18	19	20
5	350	710	710	710	710	713	715	718	718	724	730	·	·	·	·
10	175	710	710	710	713	715	718	721	724	726	730	738	752	752F	760F
15	117	710	710	713	713	718	721	724	726	730	738	752	752	760	·
20	88	710	713	713	715	718	721	726	730	732	752	752F	760	·	·
25	70	713	713	713	718	721	724	726	730	732F	·	·	·	·	·
30	58	713	713	715	718	721	724	730	732	738F	752	760	·	·	·
40	44	713	713	715	721	724	726	732	732F	752	752F	760F	·	·	·
50	35	713	715	718	721	726	730	732F	738F	752	760F	·	·	·	·
60	29	713	718	721	724	730	732	738	752	752F	·	·	·	·	·

Table 19. Torque and speed output chart for reducer of 60 and output speed of 29 rpm

Non-flanged reducers					Flanged reducers (in a gearmotor)										
Gear capacity					Ratings			Available styles							
Output torque (lb-in)	Horsepower		Efficiency	Size	Motor hp	Output (lb-in)	Service class								
	IN	OUT						F	QC	FAN	HF	SF	HQC	RF	SS
552	1.13	1.02	0.90	718-15	1	489	I	•	•		•	•	•		•
					0.75	367	II	•	•		•	•	•		•
					0.50	244	III	•	•		•	•			•
841	1.72	1.56	0.90	721-15	1.5	733	I	•	•		•	•	•		•
					1	489	II	•	•		•	•	•		•
					0.75	367	III	•	•		•	•	•		•
1,159			0.92	724-15	2	990	I	•	•		•		•		
					1.5	743	II	•	•		•		•		
					1	495	III	•	•		•		•		
1,466			0.92	726-15	3	1,466	I	•	•		•	•	•		
					2	994	II	•	•		•	•	•		•
					1.5	745	III	•	•		•	•	•		•

Figure 38. Principal stress in the x-axis

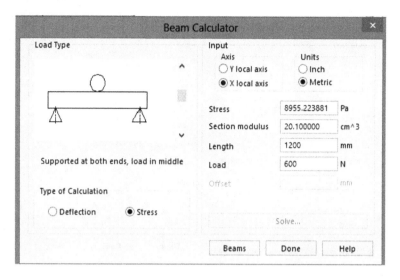

Figure 39. Principal stress in the y-axis

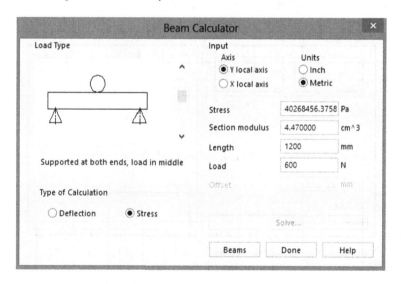

Figure 40. Deflection of the beam

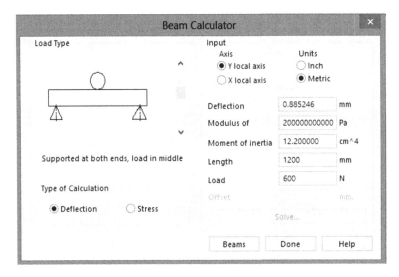

Table 20. Number of teeth on the smaller sprocket

Type of chain	Number of teeth at velocity ratio					
	1	2	3	4	5	6
Roller	31	27	25	23	21	17
Silent	40	35	31	27	23	19

Table 21. Factor of safety (n) for bush roller and silent chains

Type of chain	Pitch of chain (mm)	Speed of the sprocket pinion in r.p.m.								
		50	200	400	600	800	1000	1200	1600	2000
Bush roller chain	12 – 15	7	7.8	8.55	9.35	10.2	11	11.7	13.2	14.8
	20 – 25	7	8.2	9.35	10.3	11.7	12.9	14	16.3	–
	30 – 35	7	8.55	10.2	13.2	14.8	16.3	19.5	–	–
Silent chain	12.7 – 15.87	20	22.2	24.4	28.7	29.0	31.0	33.4	37.8	42.0
	19.05 – 25.4	20	23.4	26.7	30.0	33.4	36.8	40.0	46.5	53.5

Figure 41. Ladder logic diagram

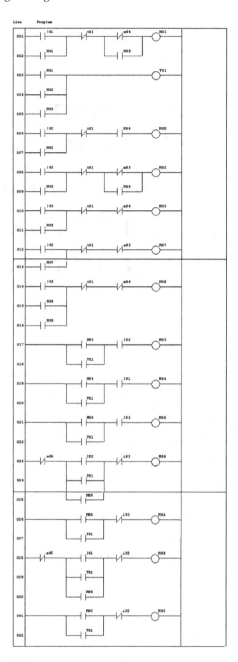

Chapter 9

Fuzzy Logic Application and Condition Monitoring of Critical Equipment in a Thermal Power Generation Company

ABSTRACT

The final case study application was at a thermal power generation company. In large-scale industrial applications, the controlling and optimization of the parameters must be done efficiently and effectively so as to attain smooth operation of the plant. In this research, the main control parameters in the boiler were identified as coal flows, temperatures in the combustion zone, air-to-fuel ratio and ash content, that is, the percentage of ash in raw coal, and mineral content. These parameters are monitored to avoid clinker formation in the super heater tubes of the boiler. Condition-based maintenance (CBM) approaches were used to monitor the boiler parameters. Fuzzy logic was applied in the monitoring of these parameters.

INTRODUCTION

Clinker formation is a serious problem in the boilers of thermal power stations which may result in forced outages for very long duration and generation loss if not attended to. Stage 2 Unit 6 boiler is forming clinkers on the platen super

DOI: 10.4018/978-1-5225-3244-6.ch009

heater elements and this has been the primary cause of platen super heater tube leaks. This has contributed to reduced boiler efficiency and increased in down time on the boiler which has impacted on the total output of the station.

CLINKER FORMATION

Clinkering is the formation of non-combustible residue that is fused into an irregular lump that remains after combustion of coal (William Collins and Co. Ltd, 2012). Clinkers are mainly a mass formed on the furnace walls due to low fusion temperature of ash present in coal these clinkers are rough and strong bonded with surface in appearance. Presence of silica, calcium oxide, magnesium oxides ad other mineral matter in ash lead to low fusion temperature. These minerals in ash differ as feed coal changes to other fused. Clinker on the furnace walls has tendency to grow and generally sticks to the hot surfaces rather than cold surfaces. Hence clinker formation and accumulation in furnace depends on quality of coal, ash fusion temperature according to the boiler operating parameter and conditions. Ash melting behaviour is a dominant parameter in generating clinkers as it follows mineral distribution of coal and existing temperature conditions in the boiler. Hence analysing temperature distribution of par bide streams in the boiler can prove to be valuable for power utilities to adopt corrective measures for clinker formation on the furnace walls.

Unit 6 has been clinkering regardless of the coal supplier. Platen super heater elements for Unit 6 (Boiler 6) have been affected by consistent clinker formation on platen super heater elements. The problem has been recurring and thus is a cause for concern. Unit 5 and 6 have similar boiler designs but Unit 6 is the one that is heavily affected. Clinker formation causes tube leaks on platen super heater elements and this result in forced outage of the unit and increased maintenance costs. For example Unit 6 went on a forced outage for four days due to tube leaks as a result of clinker formation in February 2014. This has led to a loss of a total of about 680MW which were supposed to be generated in those four outage days when the Unit was not in service. This can be translated to and quantified in monetary values totalling to a sum of $1 836 000 lost due to loss of power generation as a result of clinker formation and it cost the company $24 500 for the platen super heater elements to be maintained by a contractor Techno sphere Energy Services only for 5 days.

On Return to Service (RTS), secondary fuel which is diesel is first used for firing the boiler using the 12 burners in stage 2. It takes a total of 8 hours to fire the boiler using diesel until the required temperature of 1400oc to 1600oc is reached. When firing, the unit consumes 750 litres of diesel per hour and each litre of diesel costs \$1.35. Under normal operating conditions when the Unit is in service without clinkering, the 170MW generation will pump in \$1 836 000 into the company and no cost of over consumption of diesel will be incurred by the company since the company will be using diesel according to the planned diesel usage in which diesel will only be used for flame stabilization only for at least 1.5 hours and this translates to a cost of $8 \times 1.5 \times 750 \times \$1.35 = \$12\ 150$. This also indicates that a total of 9 tonnes of diesel ($750L \times 1,5 \times 8 = 9000L$) will be used when the unit is in service under normal operating conditions without going on an outage due to clinkering. When a unit goes on an outage as a result of clinker formation or trips, the losses that occur are:

STATEMENT OF PROBLEM

Formation of clinkers on platen super heater tubes of stage 2 unit 6 boiler, is a serious problem causing platen super heater tube leaks, poor heat transfer to the steam which then results in forced outage, increase in down time of the boiler, and generation losses. The main aim of the project is to find out the causes of clinker formation on the unit 6 platen superheaters and to come up with an intelligent CBM program for monitoring the boiler parameters. The objectives of this research are:

- To come up with an intelligent CBM program and to recommend ways to avoid clinker formation
- To find causes of clinker formation on platen super heater elements and come up with solutions to reduce the number of downtime hours by more than 240hours per year
- Reduce Unplanned Outages by 50%.

HOW THE POWER STATION WORKS

- Coal is carried either directly from the colliery on conveyer belts or from the coal stock yard and discharged into the boiler bunkers

- From the bunkers it is fed by the belt feeders into the pulverising mills where it is crushed to a fine powder by rollers on a rotating table for stage 1 or in a ball mill for stage two.
- The pulverized coal is then carried by hot air, blown through the mills by the primary air fans through ducting, into the boiler combustion chambers where it burns.
- On start-up of boiler, the pulverized coal is ignited by oil or gas burners
- Warm air is drawn from the top of the boiler house by forced draught fans and passes through an air heater into the combustion chamber for supporting combustion. Part of this air going to the combustion chamber is drawn off and blown by the primary fans through the mills to convey the pulverized coal to the combustion chamber.
- The combustion chamber walls are constructed of membrane wall tubes. Water heated in these tubes passes to the water and steam drum where steam is separated and then travels to the super heater where its temperature and heat energy are raised. From there steam is supplied to the turbine through interconnecting pipe work. The steam has a

Figure 1. Conveyor system at the power station

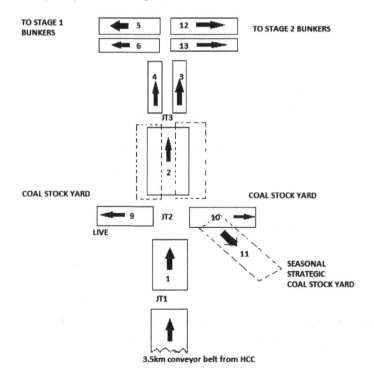

pressure of 8.9Mpa and a temperature of 518 degrees Celsius. The energy in the steam, passing through the turbine blades, causes the turbine to rotate (controlled by its governor at 3000 rpm).

- After passing though the high pressure cylinder of the turbine, steam is reheated and passed through the intermediate and low pressure cylinders. (The reheaters are on Stage 2 only).
- In its passing through the turbine blades, heat energy of the steam is converted to mechanical energy. This rotates the turbine which is directly coupled to the generator and so produces electrical energy.
- The turbine is coupled to the generator, the rotor of which is a large electromagnet whose rotation produces an electric current in the copper winding of the stator. This electric current is fed to the national grid through the transformer which increases the voltage of the electricity produced to 330kv (to minimise power losses during transmission $[I^2R]$)
- After passing through the steam, now at low pressure and temperature, enters the condenser where it is condensed back into water and eventually pumped into the boiler again by the boiler feed pump. The steam is condensed as it passes over large numbers of tubes through which cold circulating water is flowing. Incondensable gases are removed by air ejectors.

Boiler Plant

The stage two boilers (All two units) are of a two pass natural circulation reheat type capable of an evaporation of 714000 kg/hr of steam at 16.6Mpa and 543 °C at a continuous maximum rating when burning pulverised fuel, at the same time reheating to 543°C, 630 000 kg/hr of steam supplied to the reheater at 4.44Mpa and 357°C.The burners are arranged for horizontal wall firing with 2 rows on one wall and another row opposite to the wall of the two rows.

The boiler operates under balanced draft conditions maintained by forced draught fans supplying air for combustion to the burner windboxes, and induced draught fans exhausting flue gases to the chimney through rotary regenerative air heaters and an electrostatic precipitator .hot air injection from the rotary air heater discharge to the furnace controls the reheat steam temperature. Primary air fans form part of the milling plant together with coal feeders to feed raw coal to their associated pulverised fuel (PF) tube mills.

Table 1. Aspects and continuous processing

Aspects	Continuous Processing	TPDC
Products and Markets		
Product type	Standard	
Product range	Very narrow	1 Type
Customer order size	Very large	
New product Introductions	Very low	None
What is the company selling?	Product	
How are orders won?	Price	Price and reliability
How are orders won? (Order qualifiers)	Quality/delivery reliability	
Manufacturing		
Process Technology Process flexibility Production volumes Key Manufacturing Task	Highly dedicated Inflexible Very High Production	
Investment and Cost		
Capital Investment	Very high	
Inventory Levels		
Raw materials	Planned with stock	
Work In Progress (wip)	Low	
Finished goods	High	The product is simultaneously produced and consumed.
% Age of Total Costs		
Direct labour Direct Materials	Very Low Very High	
Infrastructure		
Appropriate Organizational Control	Centralized	
Appropriate Organizational Style	Bureaucratic	
Level of specialist support to manufacturing	Very High	

Fuel oil is transferred from a road and rail unloading station to the storage tanks by unloading pumps; pressure pumps draw oil from the storage tanks and delivers to the boiler which is equipped with lighting up oil burners.

The direct pulverized fuel firing boiler consists of four coal feeders, two tubes mills and their associated classifier, two seal air fans, two primary air fans and p.f. piping serving 12 burners in 3 rows. The coal feeders deliver raw coal from a bunker to a single mill and have variable output which is regulated by the mill level control system to give the desired feed of coal to the mill and boiler. The coal falls by gravity from the two feeders and enters the mill at both ends, where it is conveyed by air to the grinding elements. The coal is pulverized till the required fineness is achieved before it is air swept to the discharge boxes. Three pipes at each end of the mill serve three burners. Classifying of the pulverized coal is carried by a win-cone classifier with oversize coal particles being returned to the mill. The primary air fan delivers a controllable amount of high pressure air to the mill. The two main functions of the air are two transport the pulverized coal from the mill to the burners and to dry the coal. Hot and cold air is ducted separately to the fan inlet and by regulating the ratio the temperature of the air discharged by the fan can be controlled at a level dictated by the moisture content of the coal. With the use of primary air the mill is pressurized when operating and consequently some shaft and bearing locations could be subject to primary and pulverized coal leakage. To prevent this leakage the places concerned are sealed externally with cold air obtained from the seal air fan. To suit the normal operating range of the mill /primary air fan, the pulverized piping is designed to give velocities which at the lowest operating level will prevent pulverized coal falling out of suspension and at the highest operating level prevent undue internal erosion of the piping. The piping is also designed to be flexible as it has to absorb movements created by the boiler expansion .The pulverized fuel burners receive the primary air/pulverized fuel mixture and with integral secondary air arrangements are designed such that they inject coal and air ratio in the correct velocity range to obtain a suitable degree of turbulence of stable and efficient combustion.

Major Plant Equipment

Major items of the boiler plant include:

- One – 2pass natural circulation radiant reheat type boiler complete with a three stage superheater arranged for inter-stage attemporation. A reheater is also provided complete with emergency spray attemporators.
- One – Extended surface economizer
- One – set of long retractable soot blowers and controls
- One – set of pulverised fuel plant comprising:-
    ◦ 2 air swept ball tube mills
    ◦ 4 coal feeders
    ◦ 4 static type classifiers
    ◦ 12 circular burners

- o   2 primary air fans
- o   2 seal air fans
- One – set of oil lighting up equipment
  - o   12 'Y' jet burners
  - o   12 gas electric igniters
- One – Draught plant comprising of
  - o   2 single speed centrifugal forced draught fans
  - o   2 single speed centrifugal induced draught fans
  - o   2 rotary gas air heater
  - o   1 Dust extraction electrostatic precipitator
- One – set of structural steel work, galleries and ladders
- One – Set of insulation and refractory
- One –Set of valves, mounting and fittings
- One – Burner management system
- One – Set of Digital control and instrumentation equipment
- Common Equipment for the power station comprises:-
- One – Set of oil handling equipment
  - o   2 fuel oil storage tanks
  - o   2 Transfer pumps and strainers
  - o   3 Pressure pumps and strainers
  - o   2 Unloading pumps and strainers
- One –Set of interconnecting pipe work and valves
- One –Electrically operated goods/passenger lift
- One – Mobile crane and trailer
- One – Set of furnace inspection equipment
- One – Chemical monitoring system

Furnace

The furnace is water cooled, except for the roof which is steam cooled, and is of fully welded membrane construction designed to absorb radiant heat. The greater proportion of sensible and latent heat transfer is absorbed from radiant heat by the water cooled furnace walls. The furnace front, rear, and side walls are constructed of 63.5mm outer diameter medium carbon steel tubes on a 76.2mm pitch in prefabricated membrane panels, the roof being 63.5mm outer diameter tubes on a 114.3mm pitch ;the membrane panels are formed from evenly spaced tubes with the gap between closed by longitudinal metal strips of fins welded to adjacent tubes .the method of construction ensures a robust gas tight, continuous wall and accurate tube alignment giving ease of maintenance with minimum refractory. The furnace tube terminate at the top, bottom and at the intermediate positions into circular-section steel headers. All joints between furnace walls, floor, roof and arch are sealed by welding to form a gas tight seal .Additional roof sealing using sealing plates are arranged such that the furnace roof and reheater/superheater enclosing the roof are gas tight. This arrangement reduces the amount of refractory used in the roof sealing arrangement. The furnace is arranged for opposed firing and accommodates 12 circular pulverized fuel burners, 4 wide by 2 high on horizontal pitch of 2286 mm in the front wall, and 4 in line on a horizontal pitch of 2286mm in the rear wall.

Furnace Rear Wall and Front Wall

The front and rear water wall tubes in the furnace are equally spaced, each membrane panel tube on the front wall is 63.5mm o.d. by 7.1mm thick. From the vertical plane the tubes are set out to form the openings for the burners at the 15,155mm and 17,900mm levels. The furnace rear wall panels are bent to form a membrane arch nose and enclosure floor, 1 in 6 of the furnace rear wall tubes bifurcate into 50.8mm outer diameter tubes and 63.5mm outer diameter tubes from this point the 50.8mm tubes continue vertically to form the arch and enclosure floor with the remaining rear wall tubes. The lower ends of the wall tubes are connected to 4 front wall and 4 rear wall inlet headers arranged along the outside of the ash hopper, each inlet header being supplied through 3 tubes from the stubbed lower ends of large bore downcomers. Furnace front wall tubes are connected at their upper ends to 4 front wall top headers 3 riser tubes to each header connect the headers to stubs arranged along the front lower part of the drum. The enclosure exit screen is connected at the connect the header to stubs arranged along the Centre line part of the drum. The furnace hopper has a symmetrical unshielded throat which is supported with beams spanning the wall inlet tubing. The dimension across the throat is the 990.6mm between the tube centers.

Ash Hopper

The ash hopper of brick construction is located below the furnace hopper. Ashing must be carried out at regular intervals as necessary according to the ash content of coal to ensure that hopper loading is kept well within the limits .

Furnace Side Walls

Each side wall consists of 144 tubes in 4 panels, each comprising 36 tubes, 63.5mm outer diameter x 7.1mm thick. The lower ends of the tubes are connected to 4 left hand side wall and 4 right hand side wall inlet headers, each inlet header being supplied through 3 tubes from the stubbed lower ends of the bore downcomers. The upper ends of the wall and 4 right-hand side wall top headers, 3 riser tubes to each header connect the headers to stubs arranged along the lower part of the drum.

Insulation

The thermal and acoustic insulation consists of 100mm thick mineral fibre insulation (asbestos fibre).

Boiler Design Conditions

The boiler design conditions are given in Table 2.

The ash handling plant removes ash from the furnace ash hoppers hydraulically and transports it via high velocity sluiceways to an ash sump from where it is pumped via discharge pipelines to the ash disposal area (Ash Dam). Figure below shows the ash plant house. The stage 1 and 2 plants are designed to remove 100 tonnes of ash from the ash hopper of each boiler unit during a working period of 14 hours on the basis of the four boiler units steaming at one time.

Table 2. Boiler design specifications

Evaporation, kg/hr	714 000
Steam pressure at boiler outlet, MPa (gauge)	16.6
Steam temperature at boiler inlet , °C	543
Feed temperature at boiler inlet, °C	257
Air temperature at forced draught fan inlet, °C	35
Relative humidity %	45
Reheat steam flow kg/hr	630 000
Reheat inlet steam pressure , MPa (gauge)	4.44
Reheat inlet steam temperature , °C	357
Reheat outlet steam temperature, °C	543

The high velocity sluiceway feeds the ash to clinker grinder mounted in a pit within the ash plant house, where it is ground to a suitable size before being deposited into the ash sump below the grinder. Two clinker grinders are provided, one duty and one standby. The sluiceways are arranged so that ash may be fed to either of the clinker grinders, through a selector valve. To ensure that removal of either clinker grinder for overhaul does not interfere with the overall operation of the plant a by-pass complete with grid, is provided to feed the ash directly to the sump. Two lifting liners are incorporated in the section of the sluiceway within the ash plant house for this purpose. Slurry pumps are mounted in a pit within the ash plant house, to pump the ash and water mixture from the ash sump via the discharge pipelines to ash disposal area. Three slurry pumps are provided, one duty and two standbys. Each pump is capable of dealing with the ash and water at maximum rate at which it will be fed to the sump in the course of removal of ash from the furnace hopper. There are two discharge lines to the ash disposal area & one return line from the ash dam to the water reservoir. The output from any one of the three slurry pumps may be routed to either one of the discharge pipelines as required. Each discharge pipeline is fitted with a pneumatically operated emergency isolating valve to prevent drain back from discharge piping. High pressure pumps within the ash plant house draw water via a common strainer from the reservoir adjacent to the ash plant house and supply high pressure water for operation of the plant. Three high pressure pumps are provided, one duty and two standbys. Level detectors within the ash sump control the operation of a pneumatically controlled make up valve which in turn controls the admission of makeup water to the sump .The make-up water is taken from the reservoir adjacent to the ash plant house. Two electrically driven vertical spindle bilge pumps are provided one for draining the slurry pump pit and the other for dealing with drainage in the ash sump during non-ashing periods.

The second stage ash plant consists of two more pumps designed to enable the combined first and second stage pumps to achieve a flow rate of 640 m^3/hr. with a total slurry system head of 143.3m. The second stage pumps discharge via valve cross over connections to the two outgoing pipelines feeding the ash disposal area. The original discharge lines from the first stage pumps and discharge change over valves are now connected to the second stage pumps, to form a two stage series pumping system. The arrangement of discharge selector valves after the first and second stage pumps permits any running combination of these pumps, and selection of either final discharge line. The stage 2 additional plant comprises of two Simon Warman series a heavy duty pumps, size 8/6 G.H.H, driven through v-belts at 650 R.P.M each

Figure 2. The ash plant

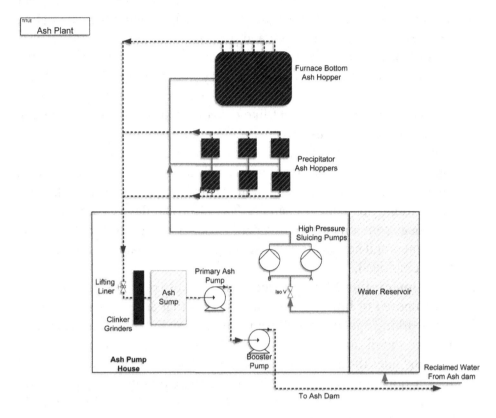

pump has full flow glands that are supplied with clean water from a gland pump mounted on the same bedplate. The gland seal pump for each ash slurry pump is fed from a common strained water supply system. However each gland pump discharge system feeds only the associated ash slurry pump. The second stage pumps are located in the stage 2 ash pump house.

PLANT MAINTENANCE

This maintenance department is responsible for the upkeep of the plant to maintain its running status and improve availability and reliability thereby increasing the plant load factor (PLF). Areas of maintenance are as follows:

1. **Civil Maintenance:** The civil maintenance in the power station covers all maintenance activities of all buildings including plant structures

and all employee residential buildings owned by the company. Other services that are in the civil maintenance include plumbing, ventilation requirements, carpentry work and painting.

2. **Mechanical Maintenance:** Mechanical maintenance refers to the maintenance activities to all machinery and equipment this includes boilers, turbines, fans , compressors, pumps, transport vehicles and material handling equipment like conveyors .lubricating and cleaning these equipments is also mechanical maintenance work.

3. **Electrical Maintenance:** This refers to all maintenance issues to pertaining to electrical equipment such as generators, transformers, switch gears motors, telephone systems, reactors, electrical installations, and lighting and battery chargers. The electrical devices are the prime logistical component to deliver the product being produced with the customers and hence this dictates that good maintenance strategies have to be employed.

4. **Instrumentation Maintenance (Process Control):** The power station is composed of intricate control strategies that are responsible for the safe working operation of all equipment. All maintenance activities that involve the digital control system (DCS) falls under process control. This includes the maintenance to the SCADA (Supervisory control and data acquisition) software, instrumentation calibration, repair as well as replacement, UPS (uninterruptable power supply) equipment, DCS servers, computers and the network structure.

Types of Maintenance Practiced at TDPC

TDPC implements the following types of maintenance:

Preventative Maintenance

Preventative Maintenance is whereby measures are taken before a failure to avoid plant unavailability and it is usually time based. An example is replacing a bearing before it shows signs of fatigue. Maximization of line performance is paramount to optimization of economic profit and thus, maintenance is a prime tool to be used to improve line performance by preserving and enhancing the function of critical assets. Knowledgeable and skilled people as well as modern tools are fundamental for the achievement of the above. The improvement of technical knowledge is therefore strongly encouraged.

Artisans receive training in strategic areas courtesy of the company, in support of the maintenance cause.

This type of maintenance also ensures reduction of occupational accidents and incidents which may result from faulty machinery.

The preventive maintenance activities in the power station are:

- Proper installations of equipment for instance proper alignment procedures, to forge readings that are out of range
- Periodic inspections of plant and equipment to prevent breakdowns before they occur.
- Repetitive servicing of equipment
- Adequate lubrication of equipment.

Predictive

Predictive maintenance (PdM) relies on the fact that the majority of failures does not occur instantaneously but develop over a period of time. By knowing which equipment needs maintenance, maintenance work can be better planned (spare parts, people etc.) and what would have been "unplanned stops" (Force outage) are transformed to shorter and fewer "planned stops"(Planned outage), thus increasing plant availability. PdM involves recording some measurements that give an indication of machine condition (temperature increase on an insulation surface, vibration increase on a bearing housing). Predictive maintenance relies heavily on instrumentation such as vibration analysers, amplitude meters, pressure, temperature and resistance transducers are used to predict trouble.

Operators who work with equipment every day can identify changes in noise levels and vibrations, temperature and so forth when doing their daily plant checks. Plant information is also relayed into the control room through the distributed digital control system (DCS), critical parameters relating to the machine part is then monitored in real-time. When important parameters deviate from the norm the (DCS) raises an alarm in the control room and thus alerts or gives a warning that something is amiss. An investigation can then be carried out to identify the exact problem. The DCS plays an important role in implementing predictive maintenance because machine parameters and unit operator actions are logged real-time giving information when a machine trips (Stops) if the problem was a human error or an equipment failure, the root cause analysis becomes easier to undertake as there is data to analyze.

Breakdown Maintenance

TPDC's maintenance regime is concentrated on breakdown maintenance because of a lot of deferred preventative maintenance activities. The power station is trying to change its operating regime so that the bulk of it maintenance jobs are preventative, this is their ideal maintenance situation. Breakdown Maintenance is simply letting the machine run its course to the point of not being able to function properly and then fix it. Breakdown maintenance results in a lot of downtime (Forced outage hours) and thereby reducing availability and the unit load factor. Examples of failures are conveyor belts being ripped, a broken shaft, a damaged bearing, a boiler tube leak.

Objectives to achieve when doing Breakdown Maintenance

- To get equipment back into operation as quickly as possible in order to minimise production loss.
- To control the cost of repair crews including regular time and overtime labour cost, this is done by interacting with the planning department to produce a valid critical path analysis.
- To control the investment in replacement spare parts that are used when machines are repaired .This calls that not only is the defective part replaced but a root cause analysis is also imitated so as to reduce future failure.
- To control the investment in spare parts and to ensure that critical parts are available in the various stores.
- To initiate opportunity maintenance so as to ensure that the appropriate amount of repairs are done at each malfunction.

Corrective Maintenance

Corrective maintenance is the repair or restoration of equipment that has a failure or is malfunctioning and not performing its intended function so that the failure is reduced. This maintenance activity is planned. The maintenance department also carries out corrective maintenance .The corrective maintenance usually consists of machine modification to curb further failure.

Plant Maintenance Plan

In order to measure energy generation performance, a reference point, or value against which to measure, is needed. For Hwange power station, this reference is the energy that could be produced per day if all the production units were operating continuously at their design output, commonly known as maximum continuous rating (MCR). However, from this gross production, a certain amount is used to drive the auxiliaries associated with each production unit. The net energy a station can produce is therefore gross output minus auxiliary power usage. All machines need to be regularly maintained and serviced. Sometimes they might suffer breakdowns. The maintenance and breakdowns, termed outages, reduce the availability of the machine and thus the net energy that can be produced. The main indicators measuring technical performance are defined as follows:

PLF-Plant Load Factor is the ratio between average load and peak load and can be expressed as

$$\mu_{plf} = \frac{P_{al}}{P_{pl}} \tag{1}$$

where

P_{al} = Average load achieved during period;
P_{rl} = Reference load

PCF-Plant Capacity Factor for a power plant is the ratio between average load and rated load for a period of time and can be expressed as

$$\mu_{pcf} = \frac{P_{al}}{P_{rl}} \tag{2}$$

where

Prl = MCR maximum continuous rating of the generator unit

Availability is the probability of a device or system being in the operating or up state at a given period t in the future. (Billinton, P.67)

$$A = \frac{St + It}{T} \tag{3}$$

where:

St = The total time the unit is in service or synchronised to the grid during the defined period.

It = Total time in idle mode because of not enough supply during the defined period, in our case It=0.

T = Defined period

Reliability is the probability of a device or system performing the purpose adequately for the period of time intended under the operating conditions encountered. (Billinton, P.59)

$$R = \frac{St}{T + Ft} \tag{4}$$

where:

Ft = Total time in failure mode during a defined period.

Analysis of the Inside of the Boiler

When the boiler was opened these are the things that were analyzed.

Figure 3 shows the bed coils in the boiler. On shut down inspection the bed coils were found to be clear of ash deposits. The coal was good with respect to slagging characteristics.

Figure 3. Diagram shows a photo of the bed coils

Figure 4 shows the FSH area. The ash deposition in the FSH area. The ash was found to bridge between the SH coils.

Figure 5 shows the fused layers of fly ash over the radiant SH placed above the furnace.

The difference in color of APH tubes confirmed the extent of air leaks? Blocks. The polishing on the casing confirmed the APH tubes had failed.

Figure 4. Diagram shows FSH area

Figure 5. Diagram shows a photo of the radiant SH placed above the combustion zone

Figure 6. Diagram shows a photo of APH tubes

FUZZY INTELLIGENCE CONDITION BASED MAINTENANCES OF CRITICAL EQUIPMENT IN A THERMAL POWER STATION

In this section a detailed analysis of the boiler parameters and how they influence the formation of clinker on the boiler, also the analysis of the ishikawa diagram for the analysis of the root causes of clinker formation on the platen super heater tubes.

Hypothesis Testing for Unit 5 and 6

A hypothesis test will be performed for both Units 5 and 6 calculating the pooled 2-sample estimate of common population variance for total air flow, feed water flow, coal flow and load, testing the difference between two means.

Airflow

Calculating the pooled 2-sample estimate of the common population variance for total airflow.

$$\sigma^2 = \frac{n_5 s_5^2 + n_6 s_6^2}{n_5 + n_6 - 2} = \frac{28\left(11.33^2\right) + 28\left(5.41^2\right)}{28 + 28 - 2} = \frac{4413.836}{54} = 81.74 \quad \sigma = 9.04$$

(5)

where:

σ = Pooled estimator for standard deviation,
n = Number of variables for unit 5,
s = Standard deviation,
μ = Mean

Testing for the Difference of Means for Unit 5 and 6

$\mathbf{H_0}$: $\mu_6 = \mu_5$ (there is no difference between unit five and unit 6)
$\mathbf{H_1}$: $\mu_6 > \mu_5$ (mean airflow for unit six is greater than that of unit five)

Consider the sampling distribution:

Table 3. Data for unit 5 and 6

Date	Unit 5				Unit 6			
	Load	Air Flow	Total Coal Flow	Feed Flow	Load	Air Flow	Total Coal Flow	Feed Flow
3/08/15	161	151	20.62	130	159	205	16.66	122
5/04/14	152	150	22.58	119	156	206	16.43	118
8/08/14	156	147	22.40	126	161	206	17.56	131
9/08/14	154	147	22.42	127	157	200	17.72	126
10/08/14	154	147	21.40	121	161	205	16.6	128
11/08/14	157	143	21.04	117	162	204	16.80	129
12/08/14	157	146	20.20	118	161	204	16.21	123
15/08/14	158	125	20.69	122	162	207	16.4	132
16/08/14	155	131	20.10	125	149	200	17.3	127
17/08/14	158	163	20.40	120	162	204	17.1	129
19/08/14	154	164	18.10	123	154	203	24.4	129
26/08/14	153	166	22.93	122	157	201	24.39	124
29/08/14	155	163	24.32	127	154	200	25.18	120
30/08/14	102	145	12.51	81	154	199	24.34	130
3/07/14	100	155	12.50	68	153	204	21.68	135
7/07/14	169	162	16.4	136	161	199	17.2	124
08/07/14	166	158	16.5	134	160	200	16.8	122
11/07/14	161	160	15.83	125	89	178	7.72	69
12/07/14	160	161	16.34	123	156	205	14.63	119
13/07/14	172	158	16.54	136	160	205	15.79	128
14/7/14	174	161	16.34	138	164	204	16.47	131
15/7/14	165	158	15.49	130	157	204	14.99	125
18/7/14	161	155	16.03	137	164	203	16.49	127
19/7/14	160	158	15.69	139	161	200	16.17	132
20/7/14	162	157	15.64	125	154	209	16.5	131
21/7/14	112	125	10.62	86	160	204	16.5	130
22/7/14	102	133	10.99	89	158	206	15.9	129
25/7/14	RTS							
27/7/14	160	153	21.72	121	154	207	15.9	117
MEAN	151.79	151.5	18.03	120.18	155.71	202.5	17.5	124.18
δ	20.31 20.68	11.33 11.54	3.73 3.8	17.4 17.72	13.35 13.6	5.41 5.51	3.53 3.6	11.56 11.77

$$\bar{x}_6 - \bar{x}_5 \sim N\left[\mu_6 - \mu_5, 9.04^2\left(\frac{1}{28_{(5)}} + \frac{1}{28_{(6)}}\right)\right] \tag{6}$$

Use one tailed test at 5% significance level. Reject H_0 if z is greater than 1.645

Calculating the test statistic (z)

$$z = \left[(\bar{x}_6 - \bar{x}_5) - (\mu_6 - \mu_5)\right]/\left[\tilde{\sigma}\left(\frac{1}{n_{(5)}} + \frac{1}{n_{(6)}}\right)\right] = \frac{(202.5 - 151.5) - 0}{9.04\left(\dfrac{1}{28} + \dfrac{1}{28}\right)} = 21.10$$

$$\tag{7}$$

Conclusion: since $z \geq 1.645$ we reject H_0. Hence there is sufficient evidence that the mean airflow for unit six is greater than that at unit five

Total Coal Flow

Calculating the pooled sample estimate of the common population variance for coal flow:

$$\hat{\sigma}^2 = \frac{28\left(3.73^2\right) + 28\left(3.53^2\right)}{28 + 28 - 2}$$

$$\hat{\sigma}^2 = 13.675$$

$$\sigma = 3.70$$

Testing the Difference Between Two Means

$H_0: \mu_6 = \mu_5$ (there is no difference in the values recorded for unit five and six)
$H_1: \mu_5 > \mu_6$ (there is a difference between the recorded values)

Consider the sampling distribution:

$$\overline{x}_6 - \overline{x}_5 \sim N\left(\mu_6 - \mu_5, 3.70^2\left(\frac{1}{28_{(5)}} + \frac{1}{28_{(6)}}\right)\right)$$

Use one tailed test at 5% significance level. Reject H_0 if z is greater than 1.645

Calculating the test statistic (z)

$$z = \frac{\left(\overline{x}_5 - \overline{x}_6\right) - \left(\mu_6 - \mu_5\right)}{\hat{\sigma}\left(\dfrac{1}{n_6} + \dfrac{1}{n_5}\right)}$$

$$= \frac{\left(18.03 - 17.5\right) - 0}{3.70\left(\dfrac{1}{28} + \dfrac{1}{28}\right)}$$

$$= 0.546$$

Conclusion: accept H_0 since $z < 1.645$. There is sufficient evidence at 5% that there is no difference between unit 5 and unit 6

Feed Flow

Calculating the pooled sample estimate of the common population variance for feed flow:

$$\hat{\sigma}^2 = \frac{28\left(11.56^2\right) + 28\left(17.4^2\right)}{28 + 28 - 2}$$

$$\hat{\sigma}^2 = 226.28$$

$$\sigma = 15.04$$

Testing the Difference Between Two Means

$\mathbf{H_0}$: $\mu_6 = \mu_5$ (there is no difference in the values recorded for unit five and six)
$\mathbf{H_1}$: $\mu_6 > \mu_5$ (there is a difference between the recorded values)

Consider the sampling distribution:

$$\overline{x}_6 - \overline{x}_5 \sim N\left(\mu_6 - \mu_5, 15.04^2\left(\frac{1}{28_{(5)}} + \frac{1}{28_{(6)}}\right)\right)$$

Use one tailed test at 5% significance level. Reject H_0 if z is greater than 1.645
Calculating the test statistic (z)

$$z = \frac{\left(\overline{x}_5 - \overline{x}_6\right) - \left(\mu_6 - \mu_5\right)}{\hat{\sigma}\left(\frac{1}{n_6} + \frac{1}{n_5}\right)}$$

$$= \frac{\left(124.18 - 120.18\right) - 0}{15.04\left(\frac{1}{28} + \frac{1}{28}\right)}$$

$$= 0.995$$

Conclusion: accept H_0 since $z < 1.645$. There is sufficient evidence at 5% that there is no difference between unit 5 and unit 6

Implication of Data

The inferred data shows that there is a significant difference in the total airflow for the data considered.

Non-Statistical Considerations

1. **Significance of Coal Flow Values:** The data for the coal flows show a general trend that there was more flow for unit five compared to unit six. The statistical average of the two units seems similar due to some uncharacteristically high feeds that occurred on six between 26 to 30 April. Neglecting these outliers there is a difference of 3kg/s on average between unit 5 and unit six. This translates to a difference of about 10 tons of coal per hour which is quite significant. This highlights that unit six had less coal feed but more total air which further increases excess air in six compared to five.

Ishikawa Diagram for the Root Causes of Clinker Formation

The Ishikawa diagram or the cause-and-effect diagram shows the causes of clinker formation, which is the specific event. This is to identify potential factors causing an overall effect. Each cause or imperfection is a source of variation. The sources are grouped into major categories to identify these sources of variation. Causes are categorized such as the 5M's, described in Table 4. The cause-and-effect diagram will reap the key relationships among variables, and the possible causes provide additional insight into the process behavior. The categories typically include People, machines, methods, materials, measurements, and environment

Figure 7 shows the ishikawa diagram which I used to find the root causes for clinker formation, on the diagram are the primary and secondary causes of clinker formation. The diagram is also known as the cause and effect diagram. The main problem or effect is clinker formation on the platen super heater tubes.

Solving the Primary and Secondary Causes

1. **Machines:**
    a. **Wearing of the Machine:** Wearing of the clarifiers may result in large particles of pulverised coal entering the boiler and the furnace for combustion and this causes high rates of formation of clinkers in the boiler and on the platen superheater tubes. Wearing of the clarifiers also affect the coal flows and usually causes the coal flows to vary with time which causes formation of clinkers

Table 4. 5M'S, primary, and secondary causes of clinker formation

5M's	Primary Cause	Secondary Cause
Machines	• Failure to remove ash and other deposits • Incorrect arrangement of tubes • Wearing of the machine	
People/manpower	• Lack of motivation • Lack of experienced plant operators, • Fatigue	Longer working hours
Methods	• Improper burner adjustments • Failure to comply with the standards • Poor maintenance planning • Incorrect maintenance method	
Materials	• High levels of mineral content in feed coal • High ash content in feed coal • Type of material used for making platen superheater tubes • Varying /increased coal flows	
Environment	• High pressures • Low fusion temperature of ash • High temperatures	Overfiring of the boiler
Measurements	• Poor calibration of pressure gauges • Poor calibration of temperatures sensors, e.g. pyrometers	

Figure 7. The ishikawa diagram for the root cause analysis of clinker formation

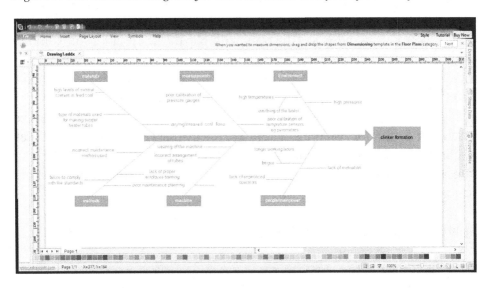

b. **Maintenance of the Clarifiers:** Should be done regularly so as to avoid wearing and in the case that there is wearing the clarifiers must be repaired before further damage. This will maintain the coal flows at constant and correct flows also the pulverized coal particles will be maintained at their proper sizes according to the boiler design.

    c.   **Incorrect Arrangement of Tubes:** When replacing the platen super heater tubes after an outage, the tubes must be properly arranged with the correct spacing because any slight changes in their spacing will promote clinker formation.

    d.   **Failure to Remove Ash and Other Deposits:** Soot blowing to remove clinkers in the boiler specifically on the boiler walls and the platen superheater tubes should be done perfectly so that any slight buildup of clinkers will be removed

2. **People/Manpower:**

    a.   **Lack of Motivation:** Workers should be motivated especially the plant operators should be given enough motivation for their job

    b.   **Lack of Experienced Plant Operators:** The plant operators should be continuously exposed to different parts of the plant and situations in the plant

    c.   **Fatigue:** This is mainly due to longer working hours, the working hours for the plant operators should be minimized especially of those who work during the night shift.

3. **Methods:**

    a.   **Improper Burner Adjustments:** The burners should be properly adjusted so that they will be tangential to the boiler to produce a fire ball like flame and this is according to the boiler design and the type of fuel used.

    b.   **Failure to Comply With the Standards:** For every boiler type there the standards that are required like the ash content the type of boiler firing, so there is need for the plant to comply with the standards

    c.   **Poor Maintenance Planning:** There is need for proper maintenance planning especially the spacing between maintenance activities should be minimized so that no damage will occur in the plant and the boiler as a result of delayed maintenance activity.

    d.   **Incorrect Maintenance Method:** The type of maintenance should be change from preventive type of maintenance to proactive type of maintenance like intelligence CBM

4. **Materials:**

    a.   **Type of Material Used for Making Platen Superheater Tubes:** Materials used for making platen super heater tubes must be strong is corrosion resistance and creep and materials like FeCrNi can be used to make platen super heater tubes

5. **Measurements:**
    a. **Poor Calibration of Pressure Gauges and Pyrometers:** All the measuring must be properly calibribrated regularly

Fuzzy Logic (Intelligence CBM)

All the other cause (parameters) are going to be monitored using fuzzy as a branch of intelligence Condition Based Maintenance.

Input Stage

The input stage is the stage that maps input, sensor such as thumbwheels, switches to the appropriate membership functions and the truth values.

For the boiler the input variables that affect clinker formation are:

1. Temperature in the combustion zone (TCZ)
2. Ash content in the raw feed coal (AC)
3. Mineral content in the raw feed coal (MC)
4. Air to fuel ratio (ATFR)
5. Coal flows (CF)

These are the 5 KPI'S.

Processing Stage

The processing stage invokes each appropriate rule and generates a result for each rule. It then combines the results of the rules to come up with the final rule.

Output Stage

This stage called the output stage converts the combined result back into a specific control output value.

To measure the rate of clinker formation on the platen superheater tubes is very difficult, so in the boiler we have three parameters which indicates the formation and the extent of clinker formation in the boiler and these are the outputs on the fuzzy system

1. Furnace Draught Pressure (FDP)
2. Heat Transfer in the Platten Superheaters (HTPSH)
3. Electrical/ Power consumption by the Induced Draught Fans and Forced Draught Fans (EC)

Figure 8 shows the boiler and the input and outputs in relation to clinker formation and also the controller.

The inputs to the boiler are monitored using the fuzzy controller.

RULES

Figures 9-11 show the variation of ash content and furnace draught pressure. It shows how the ash content affect the furnace draught pressure hence formation of clinker on the platen super heater tubes. From the fuzzy diagrams, the furnace draught pressure is at zero when the ash content is at zero but practically this is not possible for the ash content to be zero in the raw feed

Figure 8. Inputs and outputs and the fuzzy controller

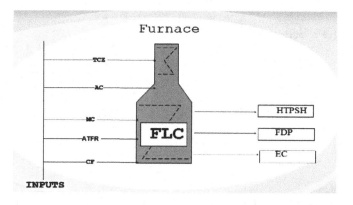

Table 5. Temperatures in the TCZ

Input 1: Temperature in the Fire Ball Combustion Zone (TCZ) [T°C]		
Low	Acceptable	Very High
0-1200	1200-1400	1400-1600

Table 6. Ash content in AC

Input 2: Ash Content in Raw Coal (% of Ash in Raw Coal) [AC]		
Very Low	**Acceptable**	**Very High**
0-20	20-30	30-100

Table 7. Furnace draft pressure

Output 1: Furnace Draught Pressure (kPa) [FDP]		
Very Low	**Acceptable**	**High**
-10-(-5)	-5-5	5-10

Figure 9. Rules from fuzzy system

1. If (TCZ is LOW) and (AC is VERY__LOW) then (FDP is VERY__LOW) (1)
2. If (TCZ is LOW) and (AC is ACCEPTABLE) then (FDP is VERY__LOW) (1)
3. If (TCZ is LOW) and (AC is VERY__HIGH) then (FDP is ACCEPTABLE) (1)
4. If (TCZ is ACCEPTABLE) and (AC is VERY__LOW) then (FDP is VERY__LOW) (1)
5. If (TCZ is ACCEPTABLE) and (AC is VERY__HIGH) then (FDP is VERY__HIGH) (1)
6. If (TCZ is VERY__HIGH) and (AC is VERY__LOW) then (FDP is VERY__LOW) (1)
7. If (TCZ is VERY__HIGH) and (AC is ACCEPTABLE) then (FDP is ACCEPTABLE) (1)
8. If (TCZ is VERY__HIGH) and (AC is VERY__HIGH) then (FDP is VERY__HIGH) (1)

Figure 10. 3D diagram of TCZ and AC vs. FDP

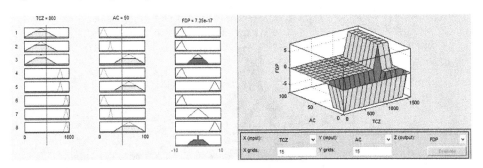

coal. The acceptable range of ash content from the diagram is from 10% to 30% but this will mean that the boiler has to operate at a negative pressure of about -7.5Kpa which is safe if the boiler (furnace) operates at a negative pressure. The boiler (furnace) is not allowed to operate with a positive pressure for and value of the ash content.

Figure 11. Graph of AC vs. FDP and TCZ vs. FDP

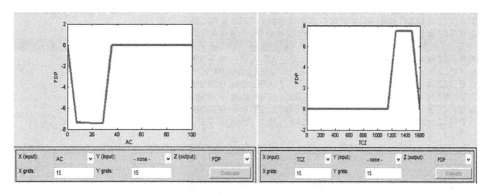

Also the diagram shows the variation of Temperature in the combustion zone and furnace draught pressure. The boiler (furnace) will operate at a pressure of 0kpa for the temperature range from 0⁰c to 1200⁰c but in this range the fuel will not undergo complete combustion and also the temperature of the superheated steam will not reach the desired temperature. The only range that is good for the complete combustion and to attain the desired temperatures of steam the temperature in the combustion zone must range from 1200⁰c to 1500⁰c but this will mean that the boiler (furnace) will operate at a positive pressure of about 7.5kpa which is not safe.

RULES

Figures 12-14 show the variation of temperature in the combustion zone and heat transfer on the platen super heater tubes. The diagram shows that the heat transfer will be at the desired values for the range of TCZ from 200⁰c to 1300⁰c meaning to say there is no effect on clinker formation as long as the temperature in the combustion zone is within this range. When the temperature in the combustion zone is within the range 1300⁰c to 1500⁰c the heat transfer is very low, about 40⁰c and this means that at high temperatures

Table 8. Heat transfer in the platen superheater

Output 2: Heat Transfer in the Platen Super Heater Tubes (T⁰C) [HTPSH]		
Very Low	Acceptable	High
0-78	78-84	84-90

273

Figure 12. Rules in the TCZ

1. If (TCZ is LOW) and (AC is VERY__LOW) then (HTPSH is ACCEPTABLE) (1)
2. If (TCZ is LOW) and (AC is ACCEPTABLE) then (HTPSH is ACCEPTABLE) (1)
3. If (TCZ is LOW) and (AC is VERY__HIGH) then (HTPSH is ACCEPTABLE) (1)
4. If (TCZ is ACCEPTABLE) and (AC is VERY__LOW) then (HTPSH is ACCEPTABLE) (1)
5. If (TCZ is ACCEPTABLE) and (AC is ACCEPTABLE) then (HTPSH is ACCEPTABLE) (1)
6. If (TCZ is ACCEPTABLE) and (AC is VERY__HIGH) then (HTPSH is VERY__LOW) (1)
7. If (TCZ is VERY__HIGH) and (AC is VERY__LOW) then (HTPSH is ACCEPTABLE) (1)
8. If (TCZ is VERY__HIGH) and (AC is ACCEPTABLE) then (HTPSH is VERY__HIGH) (1)
9. If (TCZ is VERY__HIGH) and (AC is VERY__HIGH) then (HTPSH is VERY__LOW) (1)

Figure 13. 3D diagram for TCZ and AC vs. HTPSH

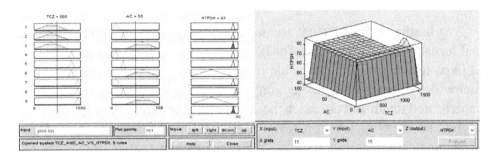

above 1500⁰c in the combustion zone there is clinker formation on platen super heaters and this means that temperatures in the combustion zone must be limited to 1300⁰c. Also the diagram shows the variation of ash content and heat transfer in the platen super heaters. The ash content does not affect the heat transfer on the platen super heater tubes in relation to clinker formation.

Figure 14. TCZ vs. HTPSH and AC vs. HTPSH

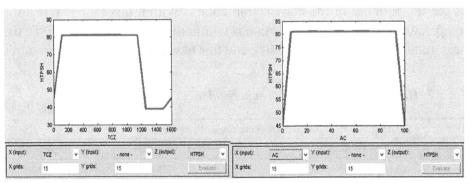

RULES FOR ID FANS

Figures 15-17 show the variation of TCZ and the electrical consumption of the ID fans. When the temperature in the combustion zone is below 1200°c the electrical consumption is 930kw for the ID fans which shows that there will be no clinker formation on the platen super heater tubes but above 1200°c the power consumption will be around 937.5 kw meaning to say that there is clinker formation and hence the ID fans has do more work and this increase in power consumption shows the formation of clinkers. Also the ash content when the percentage of ash content on the raw feed coal is above 30% there

Table 9. Electrical consumption

Output 3: Electrical Consumption of ID and FD Fans (KW) [EC]			
	Very Low	**Acceptable**	**High**
ID FANS	920-925	925-935	935-940
FD FANS	640-645	645-655	655-660

Figure 15. Rules for the ID fans

> 1. If (TCZ is LOW) and (AC is VERY__LOW) then (EC is VERY__LOW) (1)
> 2. If (TCZ is LOW) and (AC is ACCEPTABLE) then (EC is ACCEPTABLE) (1)
> 3. If (TCZ is LOW) and (AC is VERY__HIGH) then (EC is ACCEPTABLE) (1)
> 4. If (TCZ is ACCEPTABLE) and (AC is VERY__LOW) then (EC is ACCEPTABLE) (1)
> 5. If (TCZ is ACCEPTABLE) and (AC is ACCEPTABLE) then (EC is ACCEPTABLE) (1)
> 6. If (TCZ is ACCEPTABLE) and (AC is VERY__HIGH) then (EC is HIGH) (1)
> 7. If (TCZ is VERY__HIGH) and (AC is VERY__LOW) then (EC is VERY__LOW) (1)
> 8. If (TCZ is VERY__HIGH) and (AC is ACCEPTABLE) then (EC is ACCEPTABLE) (1)
> 9. If (TCZ is VERY__HIGH) and (AC is VERY__HIGH) then (EC is HIGH) (1)

Figure 16. 3D diagram of TCZ and AC vs. EC

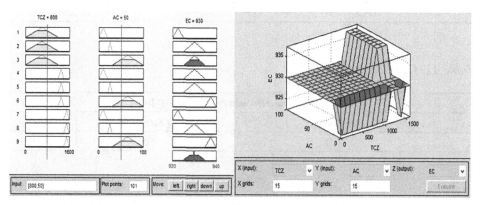

Figure 17. TCZ vs. EC and AC vs. EC

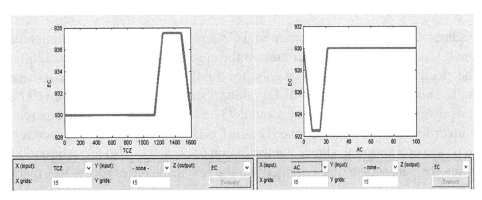

is no variation in electrical consumption. When the ash content is maintained within the range from 10-15% the power consumption will be very low and hence the variation of power consumption and ash content does not show any significance effect on clinker formation.

Table 10. Temperatures in the fire ball

Input 1: Temperature in the Fire Ball Combustion Zone (TCZ) [T°C]		
Low	Acceptable	Very High
0-1200	1200-1400	1400-1600

Table 11. Coal flows

Input 2: Coal Flows (KG/S) [CF]		
Very Low	Acceptable	Very High
0-1.8	1.8-2.5	2.5-3

Table 12. Furnace draft

Output 1: Furnace Draught Pressure (kPa) [FDP]		
Very Low	Acceptable	High
-10-(-5)	-5-5	5-10

Figure 18. Rules in the furnace

1. If (TCZ is LOW) and (CF is VERY___LOW) then (FDP is VERY__LOW) (1)
2. If (TCZ is LOW) and (CF is ACCEPTABLE) then (FDP is ACCEPTABLE) (1)
3. If (TCZ is LOW) and (CF is VERY__HIGH) then (FDP is ACCEPTABLE) (1)
4. If (TCZ is ACCEPTABLE) and (CF is VERY___LOW) then (FDP is ACCEPTABLE) (1)
5. If (TCZ is ACCEPTABLE) and (CF is ACCEPTABLE) then (FDP is ACCEPTABLE) (1)
6. If (TCZ is ACCEPTABLE) and (CF is VERY__HIGH) then (FDP is HIGH) (1)
7. If (TCZ is VERY__HIGH) and (CF is VERY___LOW) then (FDP is ACCEPTABLE) (1)
8. If (TCZ is VERY__HIGH) and (CF is ACCEPTABLE) then (FDP is ACCEPTABLE) (1)
9. If (TCZ is VERY__HIGH) and (CF is VERY__HIGH) then (FDP is HIGH) (1)

RULES

Figures 19-20 show the variation of TCZ and FDP. The diagram shows when the temperature in the combustion zone is in the range 200^{0}c to 1200^{0}c the forced draught pressure will be at -7.5kpa and that is the boiler will be

Figure 19. 3D diagram of TCZ and CF vs. FDP

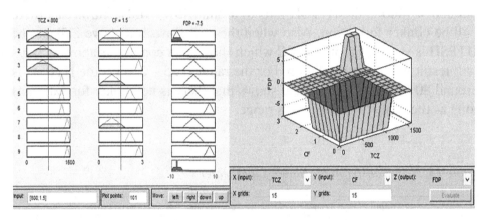

Figure 20. TCZ vs. FDP and CF vs. FDP

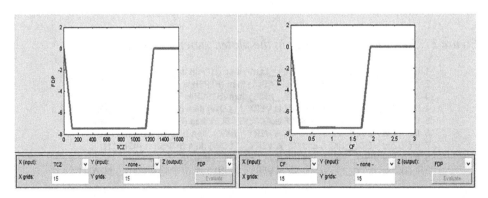

277

operating at a negative pressure which is safe and shows that there is clinker formation in the boiler. The boiler will operate at 0kpa if the temperature in the combustion zone is above 1200°c but above this temperature clinker will start to build up on the platen superheater tubes. Also if the coal flow is within the range 0.25 to 1.8kg/s the boiler (furnace) will be operating at negative pressure of about -7.5 kPa which is safe and shows that there is no clinker formation in the boiler. At coal flows above 2kg/s the boiler will operate at pressure of 0kpa and this shows that above this value of coal flows there will be clinker formation on the platen super heaters.

RULES

Figures 21-23 show the variation of TCZ and HTPSH. When the TCZ is in the range 200°C to 1400°C the HTPSH is around 80°C which is acceptable and this shows that there no clinker formation on the platen superheater tubes but above 1400°C the HTPSH falls to around 45°C which shows that there will be clinker formation. Also when the coal flows are above 2.5kg/s the HTPSH is very low around 40°C which shows that coal flows above 2.5kg/s will result in clinker formation. For the range 0.25 – 2.5kg/s the HTPSH is around 80°C which is good and shows that there is no clinker formation as long as the coal flows are in this range.

Table 13. Heat transfer in the platen superheater

Output 2: Heat Transfer in the Platen Super Heater Tubes (T°C) [HTPSH]		
Very Low	Acceptable	High
0-78	78-84	84-90

Figure 21. Rules for heat transfer in the platen superheater

1. If (TCZ is LOW) and (CF is VERY___LOW) then (HTPSH is ACCEPTABLE) (1)
2. If (TCZ is LOW) and (CF is ACCEPTABLE) then (HTPSH is ACCEPTABLE) (1)
3. If (TCZ is LOW) and (CF is VERY__HIGH) then (HTPSH is VERY__LOW) (1)
4. If (TCZ is ACCEPTABLE) and (CF is VERY___LOW) then (HTPSH is ACCEPTABLE) (1)
5. If (TCZ is ACCEPTABLE) and (CF is ACCEPTABLE) then (HTPSH is ACCEPTABLE) (1)
6. If (TCZ is ACCEPTABLE) and (CF is VERY__HIGH) then (HTPSH is HIGH) (1)
7. If (TCZ is VERY__HIGH) and (CF is VERY___LOW) then (HTPSH is HIGH) (1)
8. If (TCZ is VERY__HIGH) and (CF is ACCEPTABLE) then (HTPSH is HIGH) (1)
9. If (TCZ is VERY__HIGH) and (CF is VERY__HIGH) then (HTPSH is VERY__LOW) (1)

Figure 22. 3D diagram of TCZ and HTPSH vs. CF

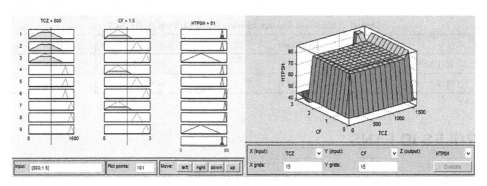

Figure 23. TCZ vs. HTPSH and CF vs. HTPSH

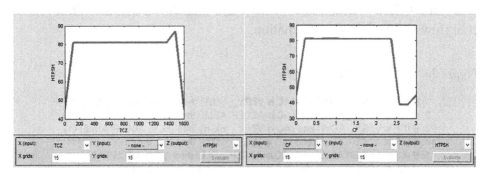

Table 14. Coal flows

Input 1: Coal Flows (KG/S) [CF]		
Very Low	Acceptable	Very High
0-1.8	1.8-2.5	2.5-3

Table 15. Ash content

Input 2: Ash Content in Raw Coal (% of Ash in Raw Coal) [AC]		
Very Low	Acceptable	Very High
0-20	20-30	30-100

Table 16. Electrical consumption in fans

Output 2: Electrical Consumption of ID and FD Fans (KW) [EC]			
	Very Low	**acceptable**	**High**
ID FANS	920-925	925-935	935-940
FD FANS	640-645	645-655	655-660

RULES ID FANS

Figures 24-26 the variation of CF and EC. When the CF is within the range 0-2kg/s the electrical consumption is around 930kw and there is no clinker formation. When the coal flow is above 2kg/s then the electrical consumption is around 937.5kw which is very high and this shows that at high levels of coal flows there is clinker formation.

Figure 24. Rules for ID fans

1. If (CF is VERY__LOW) and (AC is VERY__LOW) then (EC is VERY__LOW) (1)
2. If (CF is VERY__LOW) and (AC is ACCEPTABLE) then (EC is ACCEPTABLE) (1)
3. If (CF is VERY__LOW) and (AC is VERY__HIGH) then (EC is ACCEPTABLE) (1)
4. If (CF is ACCEPTABLE) and (AC is VERY__LOW) then (EC is ACCEPTABLE) (1)
5. If (CF is ACCEPTABLE) and (AC is ACCEPTABLE) then (EC is ACCEPTABLE) (1)
6. If (CF is ACCEPTABLE) and (AC is VERY__HIGH) then (EC is HIGH) (1)
7. If (CF is VERY__HIGH) and (AC is VERY__LOW) then (EC is ACCEPTABLE) (1)
8. If (CF is VERY__HIGH) and (AC is ACCEPTABLE) then (EC is ACCEPTABLE) (1)
9. If (CF is VERY__HIGH) and (AC is VERY__HIGH) then (EC is HIGH) (1)

Figure 25. 3D diagram of CF and AC vs. EC (ID fans)

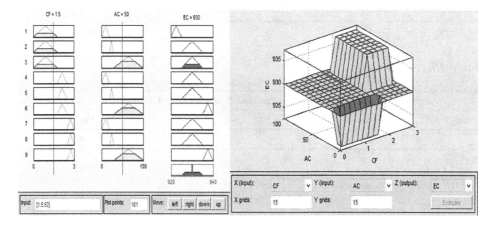

Figure 26. CF vs. EC (ID Fans) and AC vs. EC (ID fans)

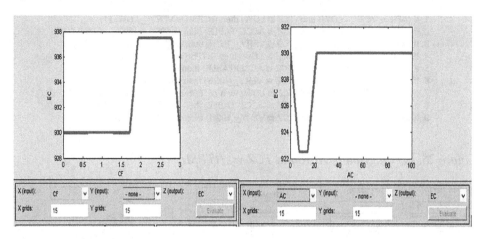

Table 17. Electrical consumption in fans

Input 1: Air to Fuel Ratio (KGs of Oxygen /KG of Coal) [AFR]		
Very Low	Acceptable	Very High
0-7	7-9	9-10

Table 18. Temperature in the fire ball

Input 2: Temperature in the Fire Ball Combustion Zone (TCZ) [T°C]		
Low	Acceptable	Very High
0-1200	1200-1400	1400-1600

Table 19. Heat transfer in the platen superheaters

Output 1: Heat Transfer in the Platen Super Heater Tubes (T°C) [HTPSH]		
Very Low	Acceptable	High
0-78	78-84	84-90

RULES

Figures 27-29 show the variation of air to fuel ratio and heat transfer on the platen super heater tubes. When the ATFR is within the range of 0.5-6.5 the HTPSH is around 39⁰c which is very low and this shows that there is clinker

Figure 27. Rules in the platen superheaters

1. If (AFR is VERY__LOW) and (TCZ is LOW) then (HTPSH is VERY__LOW) (1)
2. If (AFR is VERY__LOW) and (TCZ is ACCEPTABLE) then (HTPSH is VERY__LOW) (1)
3. If (AFR is VERY__LOW) and (TCZ is VERY__HIGH) then (HTPSH is VERY__LOW) (1)
4. If (AFR is ACCEPTABLE) and (TCZ is LOW) then (HTPSH is ACCEPTABLE) (1)
5. If (AFR is ACCEPTABLE) and (TCZ is ACCEPTABLE) then (HTPSH is ACCEPTABLE) (1)
6. If (AFR is ACCEPTABLE) and (TCZ is VERY__HIGH) then (HTPSH is HIGH) (1)
7. If (AFR is VERY__HIGH) and (TCZ is LOW) then (HTPSH is ACCEPTABLE) (1)
8. If (AFR is VERY__HIGH) and (TCZ is ACCEPTABLE) then (HTPSH is ACCEPTABLE) (1)
9. If (AFR is VERY__HIGH) and (TCZ is VERY__HIGH) then (HTPSH is HIGH) (1)

Figure 28. 3D diagram of AFR and TCZ vs. HTPSH

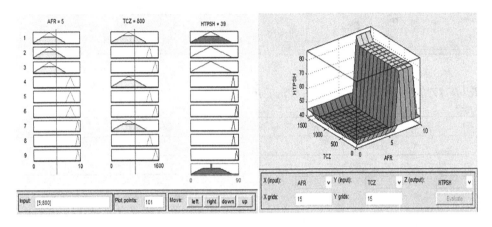

Figure 29. AFR vs. HTPSH and TCZ vs. HTPSH

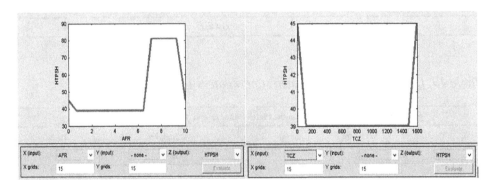

formation in the boiler which is mainly due to incomplete combustion of fuel. Then when the air to fuel ratio is above 7 the HTPSH is around 80⁰c which is the desired value this is because of the complete combustion of fuel and there is no clinker formation.

Table 20. Electrical consumption of ID fans

Output 2: Electrical Consumption of ID and FD Fans (KW) [EC]			
	Very Low	**Acceptable**	**High**
FD FANS	640-645	645-655	655-660

RULES FOR FD FANS

Figures 30-32 show the variation air to fuel ratio and electrical combustion on the FD fans. When the air to fuel ratio is in the range 1-6.5 the electrical consumption is very high up to 657.5kw which shows that there is clinker formation on the platen super heaters, when the air to fuel ratio is around 7 the electrical consumption reduces to 950kw this is mainly because there is complete combustion of fuel hence there is no clinker formation in the boiler.

Figure 30. Rules for FD fans

1. If (AFR is VERY__LOW) and (TCZ is LOW) then (EC is HIGH) (1)
2. If (AFR is VERY__LOW) and (TCZ is ACCEPTABLE) then (EC is HIGH) (1)
3. If (AFR is VERY__LOW) and (TCZ is VERY__HIGH) then (EC is HIGH) (1)
4. If (AFR is ACCEPTABLE) and (TCZ is LOW) then (EC is ACCEPTABLE) (1)
5. If (AFR is ACCEPTABLE) and (TCZ is ACCEPTABLE) then (EC is ACCEPTABLE) (1)
6. If (AFR is ACCEPTABLE) and (TCZ is VERY__HIGH) then (EC is ACCEPTABLE) (1)
7. If (AFR is VERY_HIGH) and (TCZ is LOW) then (EC is ACCEPTABLE) (1)
8. If (AFR is VERY_HIGH) and (TCZ is ACCEPTABLE) then (EC is ACCEPTABLE) (1)
9. If (AFR is VERY_HIGH) and (TCZ is VERY__HIGH) then (EC is ACCEPTABLE) (1)

Figure 31. 3D Diagram of AFR and TCZ vs. EC (FD fans)

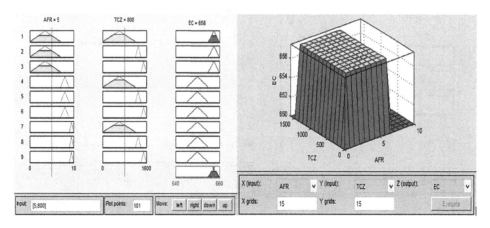

Figure 32. AFR vs. EC (FD fans) and TCZ vs. EC (FD fans)

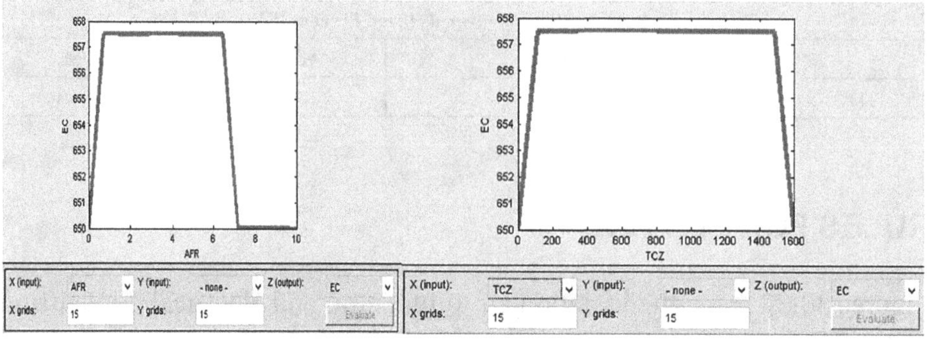

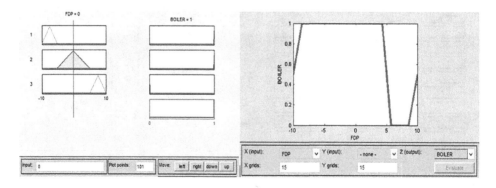

Rules for FDP, EC, and HTPSH vs. Boiler

1. **Rules (FDP vs. Boiler):** Figures 33 and 34 show the decisions that will be made either the boiler should stop or continue to run depending on the furnace draught pressure. When the FDP is within the range -7.5 – 5kpa the fuzzy controller send a message to continue run the boiler, fro -10kpa to -7.5kpa the fuzzy controller send a message to stop the boiler, also from 6kpa to 10kpa the fuzzy will send a message to stop the

Figure 33. Rules for FDP vs. boiler

1. If (FDP is VERY__LOW) then (BOILER is RUN) (1)
2. If (FDP is ACCEPTABLE) then (BOILER is RUN) (1)
3. If (FDP is HIGH) then (BOILER is STOP) (1)

Figure 34. FDP vs. boiler (either to run to stop)

boiler. The boiler will only run when the logic is 1 and when the logic is zero the boiler should stop but in the case that the logic is 0 but there is no clinker that is forming the run command will be executed that is the fuzzy read the environment and act according to the environment.

2. **Rules (EC ID and EC FD vs Boiler):** Figures 35-37 show weather the boiler should run or stop depending on the values of electrical

Figure 35. Rules for EC ID and EC FD boiler

1. If (EC__ID is VERY__LOW) and (EC__FD is VERY__LOW) then (BOILER is RUN) (1)
2. If (EC__ID is VERY__LOW) and (EC__FD is ACCEPTABLE) then (BOILER is RUN) (1)
3. If (EC__ID is VERY__LOW) and (EC__FD is HIGH) then (BOILER is RUN) (1)
4. If (EC__ID is ACCEPTABLE) and (EC__FD is VERY__LOW) then (BOILER is RUN) (1)
5. If (EC__ID is ACCEPTABLE) and (EC__FD is ACCEPTABLE) then (BOILER is RUN) (1)
6. If (EC__ID is ACCEPTABLE) and (EC__FD is HIGH) then (BOILER is RUN) (1)
7. If (EC__ID is HIGH) and (EC__FD is VERY__LOW) then (BOILER is RUN) (1)
8. If (EC__ID is HIGH) and (EC__FD is ACCEPTABLE) then (BOILER is RUN) (1)
9. If (EC__ID is HIGH) and (EC__FD is HIGH) then (BOILER is STOP) (1)

Figure 36. 3D diagram of EC (ID fans) and EC (FD fans) vs. boiler (run of stop)

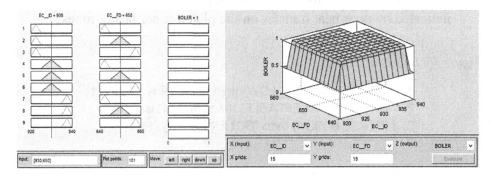

Figure 37. EC (ID fans) vs. boiler and EC (FD fans) vs. boiler

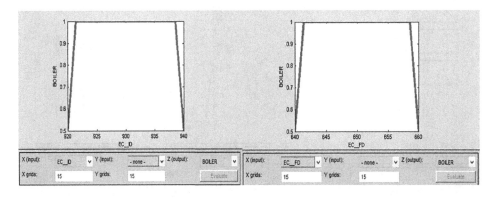

consumption. On the ID fans when the electrical consumption is within the range 922.5-937.5kw the fuzzy controller send a message to run the boiler because there will be no clinker formation and this is shown by the accepted range of electrical consumption, any value outside this range the fuzzy controller will send a message to stop the boiler. Also on the ID fans when the electrical consumption is within the range 942.5-957.5kw the fuzzy controller will send a message to run the boiler this is because there is no clinker formation, any value outside this range the fuzzy will send a message to stop the boiler because there will clinker formation on the platen super heater.

3.   **Rules (HTPSH):** The controller will send a message either to stop of run the boiler automatically. Figures 38 and 39 show the effect of HTPSH on either to stop or run the boiler. When the HTPSH is in the range 78- 83⁰C, the fuzzy controller will send a message to run the boiler and the boiler continues to run as long as the temperature range is within the range above. When HTPSH is within the range 0-77⁰C the fuzzy controller send a message to sop the boiler this is mainly because there will be clinker formation on the platen super heater tubes and is indicated by poor heat transfer on the platen super heater tubes.

Figure 38. Rules for HTPSH

> 1. If (HTPSH is VERY__LOW) then (BOILER is STOP) (1)
> 2. If (HTPSH is ACCEPTABLE) then (BOILER is RUN) (1)
> 3. If (HTPSH is HIGH) then (BOILER is RUN) (1)

Figure 39. HTPSH vs. boiler (either run or stop)

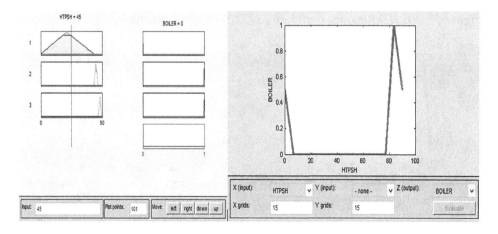

1 represents run the boiler

O represents stop the boiler

Economic Analysis

Figure 40 shows the total number of hours required for cold, warm and hot startup of the boiler, this shows that cold start up takes more hours for the boiler and the whole unit to return to its operation and this usually as a result of the boiler outage for so many hours of up to about 24hours, hot start up is usually when the unit trips and goes back to operation in a small space of time usually 2hours.

The forced outages as a result of clinker formation and hence tube leaks on the platen super heater tubes are of about 4-5 days out of operation, then the type of startup that is required is cold start up then the amount of losses that are encountered as a result of starting the boiler from outage are as a result of the fuel needed to start up the boiler. The total number of litres of diesel required to start up the boiler is proportional to the number of hours required to fire the boiler for startup. Diesel usage when bringing the Unit back in service is calculated as Total firing hours× number of firing burners × amount of diesel consumed per hour (16×8×750= 96 000Litres). On average it takes 16 hours to fire a furnace with 8 burners" hence, (16 x 8 x 750l/h x $1.35) = $129 600.00.

Figure 40. Hours taken for cold start, warm start, and hot start

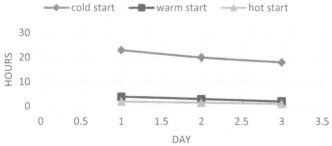

Maintenance Costs as a Result of Forced Outages on Platen Super Heaters

The maintenance costs consists of the following

- Labour costs of about $4000,
- Consumables =$500,
- Equipment costs= $20000

Which is a total of $24500 for the maintenance of the platen super heater tubes for the unit to return to service. Therefore by implementing Intelligence CBM an amount of $24500 can be saved.

Conclusion

In order to increase productivity and profitability of a company, effective maintenance has to be implemented. The fuzzy logic program showed how the different parameters affect the formation of clinker formation on the platen super heater tubes. The supplies of coal from different coal suppliers greatly affects the standard of coal required for the boiler according to the design. There were also slight variation.

Recommendations

The coal supplies from different companies varies in their coal quality that is in terms of ash content and mineral content and there is need to maintain the coal quality into the boiler from the different coal suppliers. There is

Figure 41. Hours taken for warm start and hot start up the boiler

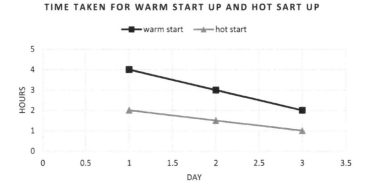

also need to implement other types of maintenance like intelligence CBM. There is also need to train the plant operators' different types of software like Matlab and their use in proactive maintenance. There is need to do the simulation for the fuzzy program so as to link the variables to the real plant data and parameters

- Energy losses due to tests may be considered as planned if they are identified at least four weeks in advance and are part of a regular program, even if the precise time of the test is not decided four weeks in advance.
- Unplanned energy losses caused by the following conditions should be included when computing the unit capability factor because they are considered to be under the control of plant management:
  ○ Unplanned maintenance outages
  ○ Unplanned outages or load reductions for testing, repair, or other plant equipment or personnel related Causes -unplanned outage extensions
  ○ Unplanned outages or load reductions that are caused by, or prolonged by, regulatory actions taken as a result of plant equipment or personnel performance, or regulatory actions applied on a generic basis to all like plants
- Energy losses due to the following causes should not be considered when computing the unit capability factor because these losses are not considered to be under the control of plant management:
  ○ Grid instability or failure.
  ○ Lack of demand (reserve shutdown, economic shutdown, or load-following)

Planned energy losses (those scheduled at least four weeks in advance) caused by the following conditions should be included when computing the unit capability factor because they are considered to be under the control of plant management:

- Planned maintenance outages
- Planned outages or load reductions for testing, repair, or other plant equipment or personnel-related causes.

Chapter 10
Fuzzy Logic and Condition Monitoring of Machinery Plant Equipment

ABSTRACT

The development of fuzzy logic in the companies mentioned by the authors is meant to automate the plant adaptively. The goal is to reduce downtimes in the plant, and hence, the framework done by the researchers will solve the problems by making the plant intelligent. Fuzzy logic is human concept, potentially applicable to a wide range of processes and tasks that require human intuition and experience. In computer, truth values are either 1 or 0, which correspond to true/false duality. In fuzzy logic, truth is the matter of degree, thus truth-values range between 1 and 0 in a continuous manner. Fuzzy logic is a method for representing information in a way that resembles natural human communication.

SCHEMATIC LAYOUT OF THE AUTOMATED PLANT

Fuzzy logic is human concept, potentially applicable to a wide range of processes and tasks that require human intuition and experience. In computer, truth values are either 1 or 0, which correspond to true/false duality. In Fuzzy logic, truth is the matter of degree, thus truth-values range between 1 and 0 in a continuous manner. Fuzzy logic is a method for representing information in a way that resembles natural human communication. It is rule based system.

DOI: 10.4018/978-1-5225-3244-6.ch010

Fuzzy logic control can be applied by means of software, dedicated controllers, or Fuzzy microprocessor embedded in digital products. Application flexibility combined with inherent simplicity and a wide range of capabilities give fuzzy logic technology a great potential growth. The researchers suggest the installation of a Supervisory Control And Data Acquisition (SCADA) with Programmable Logic Controllers (PLCs), Siemens 7 and fuzzy logic control system at the companies. Ladder logic is the preferable form of programming language for electricians and technicians responsible for the maintenance of manufacturing facilities. Since not much change in the thinking process is required to program ladder logic, it was the natural first step for the electricians who were familiar and had experience working with relays. The researchers has used this type of language. The system now operate in such a manner that there is a control room where everything that happens is noticed by the technical team. Intelligent condition based maintenance is therefore carried out in the plants.

OTHER CONTROL MEASURES FOR THE PLANT TO DO PROACTIVE MAINTENANCE

Water Control to Avoid Corrosion

Problems arising out of interference of the undesired metallic ions in the processes, the formation of scales and deposits in the equipment like heat exchangers and boilers are frequently encountered in the industry. Phosphonates offer effective solutions to many of these problems with its array of diverse key properties. These products are available in acid form as well as in the form of sodium salts, with each molecule having distinct advantage for specific applications.

Key Properties of phosphonates

1. **Sequestration:** The multivalent metal ions like Calcium, Magnesium, Copper, Zinc etc. can be complexed with Phosphonates at stoichiometric levels, forming a stable water soluble complex that suppresses unwanted effects of metal ions in various processes.

2. **Threshold Effect:** The scalant mixture can be kept in a solution form by a low concentration of Phosphonates (below the stoichiometric levels), exhibiting a threshold effect.

3. **Hydrolytic Stability:** Phosphonates are highly stable and resistant to hydrolysis over a wide range of pH and temperatures unlike polyphosphates.

4. **Deflocculation (Dispersion):** Phosphonates gets adsorbed on the growth sites of scalants. This hinders the growth of crystals from getting cluttered.

5. **Corrosion Control:** Phosphonates in combination with Zinc Phosphate, Molybdate or Nitrate, offers synergistically enhanced corrosion inhibition in water systems compared to any of the individual components.

6. **Chlorine Stability:** In the industry, Chlorine is used as an oxidizing agent as well as a biocide. Phosphonates are stable in systems containing chlorine.

HOW TO DO CBM IN THE PLANT

The plant can be monitored closely using Intelligency and can be made to illustrate what the author is emphasizing on.

WHAT'S NEXT

Suitable segments should be determined their genuine specialised qualities such as programmable rationale controllers sensors, size of scadas and different materials of specific significance for the establishment of this intelligent maintenance at the contextual investigations level. The area researched by the author is only a drop out in the sea of learning, as the world has tremendous progression in science and innovation. Without the use of a computer, it will be an extremely hard circumstance for any firm or association to make in the business sector. Choosing the correct maintenance process, just like cost optimisation, better accessibility of plants, unwavering quality and selection of high quality products is the central concern these days. For the most part, the maintenance and production procedures are done concurrently, almost 40 to 45% of cost of manufacturing can be related to the maintenance work, and consequently there is a great deal of extension to minimize the cost of maintenance.

The specialist figured out an insightful assessment framework system that utilises fuzzy logic for plant support. Aims and objectives of the writer were accomplished to recommend a superior approach to prevent incessant failures of the hardware in any production machinery. Fuzzy logic incorporated with other artificial intelligence is necessary as it might become more sensitive than the one designed here. It is necessary to combine fuzzy logic and neural network since the other non-linearity is catered for by neural network which is normally called Neuro-Fuzzy system. There is need to do this in future for machinery.

Model Reference Adaptive Control and Model Reference Adaptive Fuzzy Control need to be worked on in future in mining machinery beverages, electricity generation since they are more accurate than the general PLCs used in industries world-wide. Further research should be carried out using other techniques of AI to control valve operation and other components of the machine which contribute to the machine breakdown. These include the Genetic Algorithms (GA), Neural Networks (NN), Expert Systems (ES) and, Neuro Fuzzy (NF) systems. These techniques should each be tested on how they will respond to the bottle washer system changes and how they will be able to control the system processes. The possibility of changing the material of the valve to suit the conditions in the bottle washer should also be considered.

Due to several limitations which include time, the subject under discussion was not exhausted and there is room for future studies and improvements on the same topic. Reliability Centred Maintenance Expert systems can be integrated with other modern intelligent systems like the neural networks and genetic algorithms to help improve their efficiency as well as design towards autonomous maintenance systems which would carry out certain maintenance tasks.

MFOP is a relatively new reliability metric with different maintenance requirements from MTBF. Expert systems which use MFOP to plan maintenance strategies, scope and schedules can be rolled out. This means that expert systems would be integrated throughout equipment life cycle as opposed to independent application of different and in most cases non-compatible intelligent systems. Reliability analysis can be more easily and cheaply done by these systems. There is also need to consolidate experience from the workshop into a repository where expert systems can learn and get insight about the plant and prevent previous mistakes from recurring. The future of Expert RCM in developing nations should be rolled out not only in the power generation industry but across the other types of asset intensive industries.

For governor in conclusion the main objective of this research was to improve power generation by improving turbine response to system disturbances. A system incorporating artificial intelligence was designed and tested using Matlab and proved to be beneficial if implemented.

CONCLUSION

Presently the world has enormous advancement in science and technology the area covered by the researchers is just a drop out of an ocean of knowledge. In the present scenario if we think of life without a computer then it is very difficult for any firm or organisation to survive in the market. Higher product quality, better reliability, better availability of plants, optimisation of cost and choosing right maintenance procedure is the chief concern nowadays. Generally, the production and maintenance task are going simultaneously, nearly 40 to 45% of production cost generally goes to the maintenance work, hence there is a lot of scope to minimise the maintenance cost. The researchers managed to come up with an intelligent monitoring system framework that uses fuzzy logic for plant maintenance. Objectives and aims of the authors were achieved to suggest a better way to avoid frequent breakdowns to the equipment in any plant.

About the Authors

Tawanda Mushiri is a holder of BSc Mechanical Engineering (UZ), Master of Science in Manufacturing Systems and Operations Management (MSOM) (UZ) a D.Eng. in Engineering Management (U.J) of machinery monitoring systems in 2017. He is currently a lecturer at the University of Zimbabwe teaching Machine Dynamics, Robotics, Solid Mechanics and Finite Element Analysis. He is also the Chairman of "The Students Paperette Competition". The aim of the competition is to promote innovation and practical engineering solutions through student projects. The competition draws a single final year candidate from each of the six disciplines in the Faculty of Engineering namely Civil, Electrical, Mechanical, Geoinformatics and Surveying, Mining and Metallurgy and one other student from the Agricultural Engineering Department at the University of Zimbabwe. He is also the coordinator of Undergraduate projects and Master of Science in Manufacturing Systems and Operations Management (MSOM). Tawanda has supervised more than 100 students' undergraduate projects and 1 Masters Student to completion. He has also published 1 book, 2 chapters in a book, 9 journals in highly accredited publishers and over 60 conferences in peer reviewed publishers. He has done a lot of commercial projects at the University of Zimbabwe. He is a reviewer of 4 journals highly accredited. He has been invited as a keynote speaker in workshops and seminars. Beyond work and at a personal level, Tawanda enjoys spending time with family, travelling and watching soccer.

Charles Mbohwa is an NRF-rated established researcher and professor in the field of sustainability engineering and energy focusing on green technology, energy and systems. In January 2012 he was confirmed as an established researcher making significant contribution to the developing fields of sustainability and life cycle assessment. He has contributed a chapter to a state-of-the-art book by experts in energy efficiency. In addition, he has produced high quality body of research work on Southern Africa. Since 2012 he has worked on sustainability engineering with emphasis on integration of other soft aspects like humanitarian logistics and health care systems. The work also encompasses and integrates energy systems, life cycle assessment and bio-energy/fuel feasibility and renewable energy. He was a Japan Foundation Fellow in 1996/1997 at the International University of Japan studying technology transfer in a sustainable manner. This is the most prestigious Japanese fellowship normally reserved for postdoctoral fellows. He was a Fulbright Scholar at the Georgia Institute of Technology investigating the development of sustainable supply chains models in engineering organisations. The approaches developed have applications outside engineering. He has presented at numerous conferences and published more than 200 papers in peer-reviewed journals and conferences, 13 book chapters and two books. He holds a BSc Honours in Mechanical Engineering (University of Zimbabwe) 1986, a Masters in Operations Management and Manufacturing Systems (University of Nottingham) 1992 and a Doctor of Engineering (Tokyo Metropolitan Institute of Technology). He is a professional mechanical engineer registered with the Engineering Council of Zimbabwe and is a fellow of the Zimbabwe Institution of Engineers. He was a mechanical engineer in the National Railways of Zimbabwe from 1986 to 1991; lecturer and senior lecturer at the University of Zimbabwe and joined the University of Johannesburg as a senior lecturer in 2007. He rose to the professorship levels since then and in July 2014, was appointed as the Vice-Dean: Postgraduate Studies, Research and Innovation, Faculty of Engineering and the Built Environment. He has been a collaborator to the United Nations Development Programme, United Nations Industrial Development Organisation, United Nations Environment Programme and Visiting Exchange Professor at Federal University of Technology - Paraná, Brazil. He is an active member of the United Nations Environment Programme / Society of Environmental and Toxicology and Chemistry Life Cycle Initiative where he has served on many taskforce teams. He is one of the authors of the international guidelines on life cycle assessment databases and has been an expert reviewer on life cycle assessment methodology development. He has been invited to be a founding member of the Social Life Cycle Assessment Alliance coordinated by the World Resources Forum, which will focus, among many other things, on social impacts of sustainable recycling. He has also visited many countries on research and training engagements including the United Kingdom, Japan, German, France, the USA, Brazil, Sweden, Ghana, Nigeria, Kenya, Tanzania, Malawi, Mauritius, Austria, the Netherlands, Uganda, Namibia and Australia. His service to the research and professional communities includes being a board member of the African Roundtable in Sustainable Consumption and Production; Membership of the African Life Cycle Assessment Network: and Membership of the African Energy Policy Research Network. He has worked with the Southern African Network for Training and Research on the Environment; and was a member of the organising committee of many conferences including the International Conference on Infrastructure Development and the Environment held in Abuja and the International Federation of Operations Research Societies International Conference to be held in Johannesburg in 2008. He was listed in the Who is Who in Science and Engineering since 2011 and is an African Scientific Institute member by invitation since 2011. He is a reviewer for many journals including the following: Cleaner Production; International Journal of Life Cycle Assessment; International Journal of Humanitarian Logistics and Supply Chain Management; Energy and Fuels; Bio-resource Technology; Journal of Transport and Supply Chain Management and Environmental Science and Technology and International Journal of Engineering Research in Africa. Professor Mbohwa's fundamental work has been focused on the development of methodologies for life cycle assessment- from an environmental and economic and social point of view with interdisciplinary applications grounded in engineering. He has assisted to develop methodologies, database guidelines and has become an established international expert and reviews in this field. He has received several awards for his research papers at international conferences, including Special Commendation and several best track paper awards at the prestigious IEEE Industrial Engineering and Engineering Management conferences; Best paper and best track paper awards for World Congress on Engineering Conferences and best track and session papers for International Conference on Industrial Engineering and Operations Management. He has together with his research group contributed substantially to the research outputs of the University of Johannesburg. Between 2012 and 2014 he published 14 journal papers; 5 of journals published in had impact factor 3 and above and one of journals has impact factor 5.9. In 2015 7 journal papers have been published by his research group and one is in press. Approximately 60 conference papers have been either published or accepted. One book has been published and 4 book chapters have been accepted. Two book contracts have been signed. In total, as at 27th December 2015, his Scopus h-index was 7. He had 74 documents picking on Scopus. 11 documents in 2015; 23 documents in 2014; and 16 documents in 2013. His citations on Scopus have steadily increased from 8 for 2012 only; to 15 for 2013; 26 for 2014. There are 35 citations for 2015 and 5 in 2016. Professor Mbohwa has successfully supervised 4 postdoctoral fellows. One of the postdoctoral fellows supervised Dr. Agata Lo Giudice has attained Associate Professor status in the highly competitive Italian environment. She has published a journal paper in a 5.9 impact factor journal recently. The other Dr Anup Pradhan also published a paper in an impact factor 5.9 journal and managed to source R1.5 million external funding. The third postdoctoral fellow Dr Edoun, has published 6 journal papers in one year and one of the papers appears on Scopus already. The fourth one Dr Aboyade, has left after assisting to cement a CoJ R6 million funding programme. Through his collaboration with Professors Valerie Thomas and Bhavik Bakshi he has informally supervised 11 doctoral students at the Georgia Institute of Technology and the Ohio State University in the USA. He is an active member and doctoral students supervisor of the International Institute of Applied Systems Analysis. This is one the most prestigious research institutes worldwide having been formed to enable the collaboration of the best scientists in the Eastern and Western blocs during the Cold War era. He has successfully supervised 7 doctoral students during their summer programmes and one of them Dr. Tindall from the USA was the best student internationally in this programme. He has many UJ students completing doctoral students and one of them Mr Michael Mutingi, has been in the FEBE top ten researchers by units submitted to DHET in 2013 and in 2014. He has published one book, more than 13 book chapters, more than 45 accredited conference papers and 3 subsidised journal papers. His current Scopus h-index is 5 with 39 documents picking up on Scopus. Professor Mbohwa has successfully supervised more than 30 Master's students. All UJ Master's students have published at least one conference paper. One of them is a successful research scientist at CSIR and has published two journal papers, one of them in an impact factor 3 and above journal.

Index

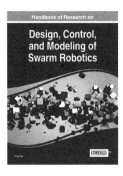

Stay Current on the Latest Emerging Research Developments

Become an IGI Global Reviewer for Authored Book Projects

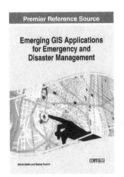

Premier Reference Source

Emerging GIS Applications for Emergency and Disaster Management

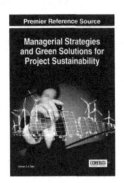

Premier Reference Source

Managerial Strategies and Green Solutions for Project Sustainability

Premier Reference Source

Comparative Approaches to Using R and Python for Statistical Data Analysis

Premier Reference Source

Solutions for High-Touch Communications in a High-Tech World

The overall success of an authored book project is dependent on quality and timely reviews.

In this competitive age of scholarly publishing, constructive and timely feedback significantly decreases the turnaround time of manuscripts from submission to acceptance, allowing the publication and discovery of progressive research at a much more expeditious rate. Several IGI Global authored book projects are currently seeking highly qualified experts in the field to fill vacancies on their respective editorial review boards:

Applications may be sent to:
development@igi-global.com

Applicants must have a doctorate (or an equivalent degree) as well as publishing and reviewing experience. Reviewers are asked to write reviews in a timely, collegial, and constructive manner. All reviewers will begin their role on an ad-hoc basis for a period of one year, and upon successful completion of this term can be considered for full editorial review board status, with the potential for a subsequent promotion to Associate Editor.

If you have a colleague that may be interested in this opportunity, we encourage you to share this information with them.